SpringerBriefs in Environmental Science

SpringerBriefs in Environmental Science present concise summaries of cutting-edge research and practical applications across a wide spectrum of environmental fields, with fast turnaround time to publication. Featuring compact volumes of 50 to 125 pages, the series covers a range of content from professional to academic. Monographs of new material are considered for the SpringerBriefs in Environmental Science series.

Typical topics might include: a timely report of state-of-the-art analytical techniques, a bridge between new research results, as published in journal articles and a contextual literature review, a snapshot of a hot or emerging topic, an in-depth case study or technical example, a presentation of core concepts that students must understand in order to make independent contributions, best practices or protocols to be followed, a series of short case studies/debates highlighting a specific angle.

SpringerBriefs in Environmental Science allow authors to present their ideas and readers to absorb them with minimal time investment. Both solicited and unsolicited manuscripts are considered for publication.

More information about this series at http://www.springer.com/series/8868

Neil Shifrin, PhD

Environmental Perspectives

A Brief Overview of Selected Topics

Springer

Neil Shifrin, PhD
Berkeley Research Group, LLC
Waltham
Massachusetts
USA

ISSN 2191-5547   ISSN 2191-5555 (electronic)
ISBN 978-3-319-06277-8   ISBN 978-3-319-06278-5 (eBook)
DOI 10.1007/978-3-319-06278-5
Springer Cham Heidelberg New York Dordrecht London

Library of Congress Control Number: 2014939308

Printed on acid-free paper

Springer is part of Springer Science+Business Media (www.springer.com)

Contents

Author Biography

Neil Shifrin has been an environmental engineer for over 40 years and specializes in water quality and hazardous wastes. He earned a B.S. in chemical engineering from the University of Pennsylvania and a Ph.D. in environmental engineering from MIT. Dr. Shifrin's early career focused on Clean Water Act issues related to Boston Harbor and the Charles River. After graduate school, Dr. Shifrin consulted on Love Canal and helped develop some of the original Superfund approaches to hazardous waste site responses. Dr. Shifrin co-founded a consulting firm specializing in human health risk assessment, contaminant transport and fate, and environmental chemistry, while consulting on hundreds of hazardous waste sites throughout the nation. Several important achievements during this consulting included the discovery of flaws in gas chromatography–mass spectrometry (GC/MS) analysis for certain contaminants, development of the BSAF (biota-sediment accumulation factor) concept for fish bioaccumulation, promotion of using an average target cleanup level for risk-based responses, and some of the earliest research on NAPL (nonaqueous phase liquids). Dr. Shifrin also has extensive knowledge of the history of waste management dating back to the 1800s. Presently, Dr. Shifrin consults mostly on multiparty allocations of response costs, the valuation of environmental project portfolios, and large river sediment contamination issues.

Chapter 1
Introduction

Abstract Environmental management has improved tremendously since the Industrial Revolution, but there is still room for improvement. Recycling continues to increase, but so does waste generation and natural resource depletion. This book is intended to provide a perspective on several pertinent environmental topics that we face today. It will serve as an introduction to students contemplating an environmental career and for individuals involved with the issues covered herein.

Keywords Resource use · Introduction · Waste · Carrying capacity

The environment nourishes our needs and assimilates our wastes. From time to time or place to place, we overwhelm the environment and struggle to repair it. From sewage filth running down medieval city streets to Industrial Revolution fish kills to climate change and natural resource limitations, environmental issues have become larger and more complex as the world's population has increased. The ecological concept of "carrying capacity" may soon rise to global consideration.

Fortunately, environmental science has progressed tremendously since the "early days" of the mid-twentieth century. In addition, environmental regulation and polluter responses, although not perfect, have made environmental protection a priority, at least in the USA. Also, natural resource limitations are recognized to varying degrees, and people are learning how to recycle some of our key resources. For example, more than 66 % of the paper and 8 % of the hazardous waste in the USA are recycled.

However, we are not yet out of the woods. Time will tell if our improving environmental science and natural resource management will be enough to address successfully the exponential population and technology growth that drives us to the brink of our carrying capacity.

Some environmental statistics for the USA that indicate the scope of our management challenges include:

- The US Geological Survey (USGS) estimates that the USA uses about 350 billion gallons of freshwater every day, with 23 % from groundwater. This is 7 % of the total precipitation in the USA. Currently, demand is slightly decreasing.

N. Shifrin, *Environmental Perspectives*, SpringerBriefs in Environmental Science, DOI 10.1007/978-3-319-06278-5_1, © The Author(s) 2014

- Publicly supplied water totals 44 billion gallons/day. It is used domestically by 258 million people (26 billion gallons/day), by industry, and for many other commercial uses; one third is from groundwater. There are 160,000 public water supplies in the USA, 84% of which serve fewer than 3,300 people each. Industry gets 18 billion gallons/day—83% from surface water—and homes get 3.8 billion gallons/day of water, mostly from groundwater.
- Agriculture (including irrigation for areas like golf courses), aquaculture, and husbandry use about 140 billion gallons/day of this freshwater.
- Thermal power plants are the largest single consumer of water, mostly for cooling, and use 200 billion gallons/day, 28% of which is saline.
- Publicly Owned Treatment Works (POTWs), which receive both domestic and commercial wastewaters, treat 32 billion gallons/day of wastewater; another 4 billion gallons/day of wastewater are disposed in the ground on site at domestic (25% of all homes), commercial, and public facilities. The USA has about 16,000 POTWs, with 2.3 million miles of sewers that treat the sewage of 75% of the population, in addition to some industrial wastewaters.
- Typical air pollutant emissions (metric tons/year) are:
  - Carbon monoxide: 57 million
  - Carbon dioxide: 5.5 billion (34 billion worldwide, with air concentrations about to reach 400 ppmv (775 mg/m^3) CO_2, up from 280 ppmv in 1800)
  - Nitrogen oxides: 13 million
  - Volatile organics: 11 million
  - Lead: 2,000 (highly improved since leaded gasoline was banned)
- The food production land requirement in the USA is about 1.2 acres per person.
- Every year, the USA mines 1.2 copper, 53 iron, and 29 phosphate rock (in million metric tons).
- The USA generates about 25 million t per year of hazardous Resources Conservation and Recovery Act (RCRA) waste and more than 250 million t per year of municipal solid waste (plus another 7.6 billion t per year of industrial nonhazardous solid waste). About 20 million of the 25 million t per year of RCRA wastes are liquids injected into deep wells—most of the rest is landfilled.
- There are 1,900 landfills in the USA (the number is decreasing, but the size is increasing) with the largest accepting 9,200 t/day. There are 21 hazardous waste landfills.
- There are about 100 nuclear reactors in the USA generating about 2,000 t of spent fuel each year; 65 of the reactors are power plants.
- The USA used 97 quads (quadrillion BTU) of energy in 2011, up from 2 quads in 1850 and about 72 quads in 1975 (the country imports 30%). The US energy supply is 36% petroleum, 26% gas, 20% coal, 9% renewables, and 8% nuclear.

As the title suggests, this book is intended to provide a perspective on several pertinent environmental topics that we face today. It is hopefully very readable and sometimes provocative to provide a big picture. It will serve as an introduction to students, who, as a result, will be in a better position to learn the details. Or it can serve as a primer on topics with which nonexperts are involved. It does not cover every important environmental topic, but it does cover several fundamental and often poorly understood issues, such as environmental sampling and analysis. The final chapter on product safety may seem out of place, but it actually reflects an important microenvironment that is currently poorly understood and poorly regulated.

Chapter 2
The Historical Context for Examining Industrial Pollution

Abstract The early focus of environmental quality was on air and surface water pollution. Land disposal and its impacts were essentially unregulated; in fact, such impacts were eventually exacerbated due to the focus on surface water protection. An understanding of groundwater contaminant transport was not developed until the 1970s and 1980s. For most of the twentieth century, pollution definitions were rudimentary and expressed in terms of "conventional pollutants," such as suspended solids, bacteria, dissolved oxygen, and nutrients. A chemical definition of pollution did not really first appear until US Environmental Protection Agency established its 64-chemical Priority Pollutant list in 1976. Early regulation of pollution was nuisance-based and enforced on a case-by-case basis using riparian rights and common law notions. By contrast, today environmental statutes and regulations provide a highly structured framework for discharge compliance and remediation of legacy contamination.

Keywords Environmental regulation · Pollution · Nuisance · Legacy contamination · Manufactured gas plants · Wastes

Introduction

Pollution today is highly regulated: Wastewater and air discharges are tightly permitted against precise numerical quality standards, and solid/chemical waste landfilling is carefully controlled. Even stormwater runoff is controlled, and even some natural background conditions are viewed as unacceptable (Mukherjee et al. 2006). Science allows us to understand pollution, technology is available to control it, and society demands as little of it as possible.

So why is there so much left? One reason is that past practices have left a legacy of contamination that we still deal with today. This is partly because pollution regulation during those past practices was very different.

As we deal with legacy pollution, we question, who should pay the bill? The Congress solved that problem with the vast liability net of Superfund, which bills anyone still standing today who touched the waste historically. Those entities fend off the bills by sending them to insurance companies, arguing that past re-

N. Shifrin, *Environmental Perspectives*, SpringerBriefs in Environmental Science, DOI 10.1007/978-3-319-06278-5_2, © The Author(s) 2014

leases were accidents and that they have insurance for accidents. Whether coverage applies depends on what was expected and intended in the past—and that depends on what scientific knowledge and regulations existed for pollution in the past.

This chapter examines the historical context for pollution views so that a "reasonable person" test might be applied to that "expected/intended" question. Although this chapter is about *all* historical industry pollution, it uses as an example, a now-forgotten but once fundamental industry—manufactured gas. The industry made gas from coal and oil, and flourished circa 1850–1950, at the end of the Industrial Revolution and just before the advent of modern environmental concerns. The manufactured gas industry represents a common duality of necessity and unwanted baggage, and it was one of the best environmental players from the Industrial Revolution. Manufactured gas represented celebrated progress every time a new plant opened. This chapter shows that even an industry as advanced as manufactured gas left a legacy of pollution, because:

- The definition of pollution was so different.
- Pollution regulation was very different.
- Pollution was viewed as temporary, because it was believed that nature would "self-purify."

To understand this historical context appropriately, this chapter considers two important concepts: objectivity and consensus. A number of objective measures of pollution exist to chronicle their changes over time, such as:

- Ability to measure pollution
- Pollution-control practices
- Scientific description of pollution mechanisms
- Laws
- Regulatory controls (e.g., discharge permits)
- Environmental quality standards

Consensus, rather than exception, is important because it represents what a reasonable person both knew and should have known. Any issue will have exceptional views, but it is only reasonable to expect someone to believe the consensus view. This is especially true with an issue of science, such as pollution, which must also pass through the scientific method before entering consensus. Some exceptional views do not pass that test, so it is not reasonable to argue that any exceptional view should be adopted (e.g., "should have been known") before it is widely accepted.

A more difficult issue straddling consensus versus exceptional views is the type of "person" in "reasonable person" category. A factory worker, a plant manager, an engineer, and a university professor might be expected to have different levels of knowledge about a particular issue, especially something as technical as pollution. As a rule of thumb, specialists (e.g., the engineer or professor) might be expected to be more aware of exceptional views, but all persons should still be expected to believe no further than the consensus view.

Perhaps the most important difference in views between the past and present is that essentially none of today's hazardous waste issues were considered or even understood to be an issue during the MGP era. This means that an MGP operator could not have expected or intended the type of damage being addressed today, because an awareness or understanding of that damage did not exist. In addition, the cause of nuisance-type impacts was poorly understood, and the practical ability to predict when such problems might occur was limited.

To the extent that impacts from pollution were understood in that former era, the issue was whether there was any tangible indication of a harmful effect, typically detectable by sight, smell, or taste. The environment was viewed as having the ability to attenuate impacts from wastes through "self-purification." The focus of the time was on surface water bodies, and land disposal of wastes was favored.

Protection of Surface Water

Consistent with historical understanding at the time, some early statutes existed that concerned discharges into surface water, but they were not universally applied. For example, an early New York statute from the Civil War era made it unlawful to "throw or deposit any gas-tar" or "refuse" of "gashouses or gas factories" into "certain waters" (New York Legislature 1845). In 1881, the law was modified to prohibit the throwing of "gas-tar, or the refuse of a gas house or gas factory, or offal, refuse, or any other noxious, offensive, or poisonous substance into any public waters" (Parker 1907). After amendments over the years, these and other early laws were amended to include caveats such as "destructive" (e.g., New York Legislature 1892; State of New York 1900) or "injurious" to fish (e.g., State of New York 1912, 1913; New York Environmental Conservation Law 1972). Such caveats reflect "reasonable use" concepts applied at the time, as discussed below.

An example of the way statutes with environmental elements were enforced during the MGP era is described by Andreen (2003) for the 1899 Rivers and Harbors Act, which is often cited as the prevailing federal water quality control law prior to the 1972 Clean Water Act (CWA). Section 13 of the Rivers and Harbors Act prohibits the "discharge of *any refuse matter of any kind, whatever*," except municipal sewage into navigable waters. [Emphasis added.] The US Army Corps of Engineers, the statutory enforcer of the act, for decades, interpreted Section 13 to apply only to the discharge of materials that could impede navigation. By this law, nuisance was interpreted as a solid material that might interfere with boat traffic. The enforcement of this law did not expand until the 1960s, when the US Supreme Court ruled that Section 13 applied more generally to industrial pollution.

Although many of these early statutes were broadly worded, history shows that these statutes were not in practice enforced to prohibit all discharges to surface waters. For example, more than 50 MGPs were observed in detail in a 1919 report by the New York State Department of Health (NYSDOH), and discharges were noted, but no laws were claimed to have been violated.

Land Disposal, Landfilling, and Their Regulation

In contrast to even mild surface water regulation, there were no comparable statutes governing the disposal of wastes on land, historically. This is not surprising, given the limited understanding of the impacts of land disposal of wastes during the MGP era. This limited understanding of contamination and the almost total focus on surface water had the unintended consequence of causing the environmental problems on land and in groundwater that are the subject of today's cleanup actions.

Through the mid-twentieth century, most regulation and literature on land disposal of waste focused on the sanitary landfilling method of municipal waste disposal, a practice which began in the 1930s (Moore 1920; Eddy 1934, 1937; Cleary 1938; Civil Engineering 1939; APWA 1966; Mantell 1975; Wilson 1977). Sanitary landfilling involved compaction and daily soil cover of the waste. It was developed to ensure efficient land use and to keep wastes covered to avoid disease vectors such as rats and insects.

The US Public Health Service (USPHS) endorsed the sanitary landfilling methods in 1940 (Phillips 1998), but it was not until 1959 that the American Society of Civil Engineers (ASCE) published technical standards for it (ASCE 1959), after most MGPs had been closed. These standards were not systematically followed: A 1973 survey indicated only about 15% of solid waste disposal in the USA followed the standards, with "almost all of the balance [being] disposed of simply by open dumping" (Baum and Parker 1973). Until the 1970s, open dumps with burning and incineration were the primary methods of solid waste "management" (Phillips 1998).

The literature on land disposal of industrial wastes was more limited than that for municipal waste during the first half of the twentieth century, but the practice was widely accepted. By the 1950s, industrial waste land disposal was addressed somewhat widely in the technical literature, but more in terms of how to do it, not whether to do it (Rudolfs et al. 1952; Stone 1953; Powell 1954; Black 1961; Rosengarten 1968; Snell and Corrough 1970; Overcash and Pal 1979). The petroleum industry, for example, issued manuals of practice that described combined treatment/ surface water discharge and land disposal options (API 1951), as did the pesticide industry (NACA 1965) and the National Safety Council in terms of generally acceptable practices for industry (Gurney and Hess 1948). As late as the 1970s, the US Environmental Protection Agency (USEPA) estimated that more than 90% of all hazardous industrial wastes were land disposed (USEPA 1977).

In 1967, the US chemical industry consisted of 2,030 manufacturing plants, generating an average of 33,000 t of process waste per year per plant (Holcombe and Kalika 1973). The vast majority of these wastes consisted of fly ash (52%), sludge (39%), and filter residue (4%), with tars and off-specification product each contributing about 2%. Land disposal was the ultimate fate of 72% of these wastes, with another 10% put in lagoons, 8% incinerated, and "other" at 10%. The majority (58%) of this land disposal was onto plant property with the other 42% onto public land.

It was not until the 1970s that a scientific understanding came into existence that would allow the recognition of the potential dangers of land disposal. In earlier decades, some reports of impacts from leaks, spills, or disposal on land existed, but these reports were anecdotal without systematic explanation. Without such predictive understanding, an MGP operator could not have expected that leaks, spills, and disposal activities would cause harm. A scientific understanding of the cause–effect relationship of subsurface contamination was required but did not begin to develop until the advent of groundwater contaminant modeling in the 1970s (Konikow and Bredehoeft 1978).

This lack of scientific knowledge about land disposal effects is described in a comprehensive 1960 MIT study on land disposal sponsored by the Federal Housing Authority (FHA). That study concluded that contemporary knowledge of groundwater contamination from industrial land disposal was "not satisfactory" in terms of understanding concepts of: (1) permissible concentrations in groundwater; (2) migration of contaminants into and through groundwater; (3) the ability of soils to attenuate the contamination; and (4) the ability to predict contamination (Stanley and Eliasson 1960). Moreover, the study gave organic chemical industry wastes (the category applicable to MGP wastes) almost the lowest priority (18 out of 20) for groundwater contamination research needs.

In 1976, growing awareness of the need to address the impacts of land disposal led the Congress to enact the first modern federal statute addressing solid and hazardous waste, the Resource Conservation and Recovery Act (RCRA). Intended for "cradle-to-grave" waste control for operating facilities, RCRA established the framework for the existing US system of managing and controlling land disposal of many industrial wastes (U.S. Congress 1976; USEPA 2001).[1]

A further turning point was the 1978 Love Canal incident, which triggered a heightened level of technical investigation and public concern, eventually leading to the enactment of the Comprehensive Environmental Response, Compensation, and Liability Act (CERCLA, aka Superfund) and similar state laws that regulate hazardous waste cleanups. Thousands of tons of chemical waste had been buried in the 1940s and 1950s in the Love Canal, located in Niagara Falls, New York (USEPA 1996). In 1978, heavy rains pushed some of this chemical waste upward and into the basements of homes and onto the ground surface, leading to a massive cleanup, control, and relocation effort (New York Times 1978; USEPA 1996). The incident served at the time to demonstrate our scientific and regulatory ignorance of the environmental and public health impacts from historical land disposal of wastes.

Partly in response to the Love Canal incident, the US Congress completed a comprehensive study of industrial waste disposal in the USA. This study (the "Eckhardt Report") found that (1) 94% of the wastes disposed on the land by the industries surveyed was dumped on site; (2) few states at the time required any regulation of such disposal; (3) the total number of hazardous waste generators was

---

[1] The Toxic Substances Control Act (TSCA), also enacted in 1976, provided USEPA with authority to address the production, importation, use, and disposal of specific chemicals, including PCBs, asbestos, radon, and lead-based paint.

estimated at 272,000, using up to 30,000 disposal sites; and (4) the CWA had been responsible for shifting a large portion of industrial pollution over to land disposal (U.S. Congress 1979). The important lesson to be learned from Love Canal and the Eckhardt Report is that, even in 1980, tremendous ignorance remained as to the full dimensions of groundwater and other problems created by historical land disposal of chemical wastes, including ignorance of the technical characterization of the sources, health threats, natural resource threats, remedies, and appropriate regulatory frameworks.

Love Canal and the Eckhardt Study were part of the growing recognition in the late 1970s that many waste disposal practices of the past had created problems for the present. Even with this recognition, it took more time to understand the problem. For example, the USEPA did not issue its first RCRA regulations until 1980, despite the law's enactment in 1976. The 4-year delay reflects the learning curve about hazardous wastes at the time. Industry did not change practices during that period because it awaited forthcoming regulations. For the first time, in 1980, through CERCLA and RCRA, individual chemical contamination was broadly regulated in terms of the potential for health effects, using risk assessment and risk-based environmental concentration standards. The transition had been made from nuisance-based responses to a more comprehensive legislative, scientific, and regulatory framework for environmental management.

The environmental laws and regulations enacted in the 1970s reflect a modern understanding that even trace amounts of certain chemicals may pose a potential human health threat many years after their release into the environment. The results of this radical change in scientific understanding are the enforcement actions brought today under Superfund.

Understanding of Pollution

Understanding of pollution was very different during the MGP era than it is today. For example, during that time, the consensus understanding of pollution in general was that:

- Waste discharges were allowable: The concept of "reasonable use" prevailed, which meant that waste discharges were only precluded if they created a nuisance.
- Pollution was temporary: Self-purification by nature was expected. Thus, legacy contamination like cleaning up hazardous waste sites could not have been anticipated.
- The definition of pollution was limited to what we call "conventional pollutants" and did not include specific chemicals.

Understanding evolved, as demonstrated by changes in the ability to measure contaminants, environmental standards, waste discharge permits, and available treatment technology.

Waste Discharges were Allowable and Nuisances Hard to Predict

In the 1950s, more than 50% of all wastewater discharges were totally untreated (NEIWPCC 1951). In the 1920s, the sewage from all six million people in New York City was discharged with no treatment (U.S. Engineer Office 1925). By contrast, most waste discharges from MGPs throughout the twentieth century were treated with what was at the time advanced technology. Waste discharges were generally allowable unless they caused a nuisance. Nuisances were judged in terms of recognizable impacts within the context of the "reasonable use" of waterways.

During the MGP era, the sanitary engineering notion of "reasonable use" was applied to define whether a waste discharge was a nuisance. The concept of reasonable use was derived from common-law notions of riparian rights. Although different in the western USA, riparian rights derive from property rights, which include the right to use, but not own, water associated with owned land.

Unfortunately, the notion of riparian rights had a circular nature that resulted in much debate (USPHS 1917). This debate was due to the fact that each riparian landowner had a right to reasonable use of a waterway for both consumption and waste discharge. Since this same right was also afforded to the next downstream landowner, too much waste discharge (or consumption) by the upstream landowner might violate the consumption rights of the downstream landowner. This conundrum was balanced by considering whether the upstream use was reasonable in terms of the downstream needs.

Local practice and the nature of a use sometimes also influenced what was "reasonable." As long as the upstream use, including waste discharges, did not cause a nuisance to downstream needs, upstream activities were considered reasonable. If a waste discharge created a downstream nuisance, such as the inability to provide water to livestock or the need to filter downstream water for subsequent use, it might be called "pollution," and a lawsuit might be filed. There was no inherent preclusion of waste discharges until modern environmental regulation was instituted.

Essentially, views regarding pollution during the MGP era were the opposite of what they are today. There was no assumption that all discharges would cause harm. Although discharges were curtailed if they caused a nuisance, there was generally no way a priori to predict if a discharge would cause a nuisance. That is because a nuisance condition depended on both the nature of the discharge in terms of the receiving water and other needs/uses of the receiving water. In practice, whether harm occurred was judged by whether there was a complaint.

Pollution was Temporary

Waste load allocation studies are a common step currently used by environmental engineers for determining waste discharge limits. These determinations are based on two concepts: one newly evolving at the time of the 1972 CWA, stream classifications, and one left over from the reasonable use period of water quality control, self-purification. The notion of self-purification was underpinned by observations that oxygen-consuming organic wastes in water would be purified by biodegradation and natural reaeration, while settleable pollution (sediment contamination) would be biodegraded and buried by clean settling solids. A reliance on dilution was also a consideration.

The 1950s and 1960s saw the first comprehensive stream studies and stream quality classifications. Stream classifications, essential for water quality management, were water quality targets based on intended use and the practicalities of local conditions. In a way, these were an evolutionary clarification of reasonable use considerations. Classifications generally ranged from "drinkable" to "waste conveyance." As noted by one of the leading sanitary engineers of the time, "disposal of wastes is one recognized best usage for waters in New York State" (Rudolfs 1952). The fact that the government recognizes several different classifications for surface waters means that all water is not expected to be pristine.

Pollution Definition Limitations

During the MGP era, pollution was defined in terms of what environmental engineers call "conventional parameters," which comprised the limits of pollution understanding until the advent of concerns with specific chemicals. These included:

- Coliform bacteria: an indicator of human waste and for waterborne disease
- Suspended solids: mostly an aesthetic issue, but sometimes simply an indicator of waste
- Dissolved oxygen (DO): important for balanced life in waters
- Biochemical oxygen demand (BOD): the result of a test that examines organic material decay that affects DO in waters
- Nutrients (e.g., nitrogen and phosphorus): affect algal growth, which can impact aesthetics and oxygen balances in water

These were the definitions of surface water quality. The definition of groundwater quality was limited to taste and odor, along with bacterial quality for groundwater used for drinking. The limited historic understanding with regard to protecting groundwater from bacterial quality is reflected by local ordinances nationwide, requiring only minor separation of cesspools from drinking water wells—most often 20–100 feet (USPHS 1913, 1915, 1917). After the passage of modern environmental laws in the 1970s, chemical-specific quality definitions for groundwater, soils, and sediments were also established. This focus on chemical pollution did not start until USEPA established its original 64 "Priority Pollutants" in 1976 (dioxin, was originally delayed and now there are now 129) as a result of a Natural Resources Defense Council (NRDC) lawsuit to enforce Section 307 ("toxic chemicals") of the 1972 CWA.

Evolution of Pollution Understanding

The nuisance-based view of pollution persisted until the passage of the 1972 CWA. The CWA changed this view by requiring (1) permits for all wastewater discharges and (2) universal wastewater treatment. The newly required levels of treatment were best practicable treatment (BPT) and best available treatment (BAT), targeted for 1977 and 1983 nationwide application, respectively. USEPA had little trouble defining BPT and BAT for municipal wastewater treatment, but the agency took decades to define them effectively for industrial wastewaters because (1) technology at the time had not been aimed at industrial wastewater and (2) industrial wastewaters were so diverse, which defied a simple solution.

The second paradigm change required by the 1972 CWA was wastewater discharge permits. USEPA or state designees issued permits under the National Pollutant Discharge Elimination System (NPDES) program starting around 1974. For the first time, a permit was required for every discharge. This contrasted with the few earlier state programs that focused primarily on sanitary protection and had not been universally applied. The NPDES agencies issued permits with quantitative limitations on the discharge of certain wastes after consideration of the receiving water quality classification and its ability to assimilate the waste. The first round of NPDES wastewater permits rarely contained chemical-specific requirements. Moreover, the permits did not and still do not require zero discharge.

Over time, but generally not until the 1970s, views on environmental quality became more sophisticated. This evolution was due to:

- *Improvements in the ability to measure pollution.* Pollution measurement changes are chronicled very clearly by the 22 editions of *Standard Methods for the Examination of Water and Wastewater*, published since 1905 jointly by the American Public Health Association, American Water Works Association, and Water Environment Federation. This chronicle demonstrates not only measurement improvements but also changing definitions of pollution.
- *Improvements in the ability to control pollution.* Textbooks, published scientific papers, and proceedings from the Purdue Conferences document changes in wastewater control technology and its practice over time. For example, James Patterson, chairman of Environmental Engineering department of the Illinois Institute of Technology, published a textbook on industrial wastewater treatment technology in 1975 focusing on "22 major industrial pollutants," the most specific of which for MGP-type contamination was "oily wastes" (Patterson 1975). The textbook offered the same technologies used by the MGP industry 50 years earlier. Many other similar examples exist in the literature. The point is twofold: (1) Wastewater treatment technology was rudimentary even as late as the 1970s and (2) the MGP industry was at the forefront of available technology.
- *Improved knowledge and societal understanding of the impacts of pollution.* Although water quality standards have existed since 1914, they remained rudimentary until the 1970s (USDOI 1968; McDermott 1973). The first water quality standards were limited to a single parameter—bacteria—consistent with the

historical understanding of the impacts from pollution noted above. Today, environmental standards for hundreds of chemicals exist for many media.

• *Increasing load of pollution in the environment* over time, due to increasing population and relative to the ability of nature to abate pollution and self-purify.

The Gas Manufacturing Trade

As noted in the introduction to this chapter, the MGP era represents the historical context for pollution in the last part of the Industrial Revolution. To complete the story against this backdrop, this section describes the industry's common practices for wastewater control, for which the industry was one of the leaders of its time.

The American Gas Association, established in 1919, and numerous local trade groups (e.g., the Western Gas Association, New England Association of Gas Engineers) provide a rich literature on the challenges of making and selling gas during the MGP era. Only a small part of this literature addressed pollution control in the industry, because the main focus of the trade groups, as with any industry during the Industrial Revolution, was in making their product.

Odor was the predominant environmental complaint about MGPs during their operation, but tar, the primary MGP by-product, is the industry's environmental legacy often at issue today. The composition of MGP tar is highly variable, but in general, it contains hundreds of chemicals, including about 10 % naphthalene and 0.1 % benzene. Tar is viscous (i.e., molasses consistency) and dense (i.e., sinks through water). Today, tar is viewed as a contaminant by many states because it serves as a source to contaminate groundwater through dissolution of its constituents, soils via chemical adsorption of its constituents and absorption, and air by volatilization of its constituents. Historically, MGPs attempted to reclaim tar from wastewater as a means of pollution control and because tar was a valuable commodity—more valuable than the plant's coal feedstock—and was used in dyes, road paving, roofing, personal care products, chemical manufacturing, and wood treating, among others.

Membership in MGP trade groups did not impart any particular knowledge about pollution control, because membership could not ensure understanding of or even exposure to information. Nevertheless, the MGP trade groups disseminated information about pollution control, such as offering a standardized design for tar separators (Sperr 1921), which was adopted by many MGPs around that time. Contemporaneous commentary about environmental impacts of MGP wastes would be a function of local conditions, however, and thus highly variable. That is why generalizations such as those posed by Willien (1920) or Hansen (1916) were not met with universal agreement within the MGP industry. Pollution knowledge was still anecdotal, not mechanistic, which led to a lack of consensus.

Actual knowledge and practice during the early MGP era were as follows:

• Many plants recovered tar from an early date. For example, some MGP records reveal tar recovery equipment before 1900.

- Although earlier sedimentation designs were used, the MGP industry standardized on a baffled chamber tar separator design around 1920, and most plants used that design thereafter.
- Additional treatment beyond tar separation was the exception rather than the rule, as noted by NYSDOH's 1919 study of 53 MGPs, which showed that only 17% used further treatment, such as filtration (Biggs 1919).
- Views of industrial pollution at the time (c. 1920s) were limited mostly to the physical impacts of oil, not on chemical impacts as viewed today. The primary negative physical impacts of oil were viewed as preventing natural degradation of sanitary sewage, interference with reaeration, and unaesthetic surface coating.

MGPs represent the best of the Industrial Revolution from an environmental perspective. But MGPs and other industries of that era have left us today with a difficult environmental problem that no one back then could have anticipated.

References

American Petroleum Institute (API) (1951) Manual on disposal of refinery wastes, Supplement 1 (Biological treatment) to Vol III Chemical wastes, 2nd edn. March
American Public Works Association (APWA) (1966) Municipal refuse disposal. Prepared for Public Administration Service, Chicago
American Society of Civil Engineers (ASCE) (1959) Sanitary landfill. Prepared by the Committee on Sanitary Landfill Practice of the Sanitary Engineering Division of the American Society of Civil Engineers
Andreen W (2003) The evolution of water pollution control in the United States—State, local, and federal efforts, 1789–1972, Part II. University of Alabama Law School
Baum B, Parker CH (1973) Solid waste disposal, Vol 1: incineration and landfill. DeBell & Richardson, Inc., Contract Research and Development, Enfield, Connecticut. Ann Arbor Science, Ann Arbor
Biggs HM (1919) New York State Department of Health, fortieth annual report. Lyon, Albany
Black RJ (1961) Refuse disposal standards and regulations. In American Public Works Association Yearbook, 1961. American Public Water Works Association, Chicago, IL, pp 233–249
Civil Engineering (1939) Technical aspects of refuse disposal: a review of developments in practice and research in 1938
Cleary EJ (1938) Landfills for refuse disposal. Eng News Rec 121:270–273. September 1
Eddy HP (1934) Can refuse collection and disposal systems be improved? Am J Public Health 24:119–121
Eddy HP (1937) Refuse disposal—a review. January
Gurney SW, Hess RW (1948) Industrial waste disposal and bibliography on chemical wastes. National Safety Council. Industrial Safety Series No. Chem. 7
Hansen P (1916) Treatment and disposal of gas house wastes. Gas Age 37:401–403
Holcombe LK, Kalika PW (1973) Solid waste management in the industrial chemical industry. Report No. EPA/530/SW-33c
Konikow LF, Bredehoeft JD (1978) Techniques of water-resources investigations of the United States Geological Survey, Chapter C2: computer model of two-dimensional solute transport and dispersion in ground water. United States Geological Survey (USGS) Publications, Washington, DC
Mantell CL (1975) Solid wastes: origin, collection, processing and disposal. Wiley, Hoboken, 1127 p

McDermott JH (1973) Federal drinking water standards—past, present and future. J Environ Eng Div 99(4):469–478. August

Moore TF (1920) Prevailing methods of garbage collection and disposal in American cities—part 2. Am City 22(6):602–608

Mukherjee A, Sengupta MK, Hossain MA, Ahamed S, Das B, Nayak B, Lodh D, Rahman MM, Chakraborti D (2006). Arsenic contamination in groundwater: a global perspective with emphasis on the Asian scenario. J Health Popul Nutr 24(2):142–163

National Agricultural Chemicals Association (NACA) (1965) Manual on waste disposal. Washington, DC. June

New England Interstate Water Pollution Control Commission (NEIWPCC) (1951) Industrial wastes in the New England interstate water pollution control compact area. New England Interstate Water Pollution Control Commission, Lowell

New York Environmental Conservation Law (1972) Polluting streams prohibited. Article 11, Title 5, Section 503

New York Legislature (1845) Laws of the State of New York passed at the sixty-eighth session of the Legislature. May 13

New York Legislature (1892) Laws of the State of New York, Vol 1, passed at the one hundred and fifteenth session of the Legislature. January 5–April 21 and April 25–April 26

New York Times (1978) Upstate waste site may endanger lives. August 2

Overcash MR, Pal D (1979) Design of land treatment systems for industrial wastes—theory and practice. Ann Arbor Science, Ann Arbor

Parker AJ (1907) Section 390, Throwing gas tar, etc., into public waters. The penal code of the State of New York being Chapter 676 of the Laws of 1881, as amended by the Laws of 1882–1907, Inclusive, with notes, forms and index

Patterson JW (1975) Wastewater treatment technology. Ann Arbor Science, Ann Arbor, 265 p

Phillips JA (1998) Managing America's solid waste. Phillips, Boulder

Powell ST (1954) Industrial wastes: some trends are evident from developments that have occurred during the last twelve months. Ind Eng Chem 46(1):113–114. January

Rosengarten GM (1968) Solid waste disposal by industry. In urban America challenges engineering, Monograph I-1, Report of the college-Industry Conference, New Orleans, Louisiana, 8–9 February 1968

Rudolfs W, Bloodgood DE, Edwards GP, Ettinger MB, Faber HA, Gehm WH, Haney PD, Harris RL, Heukelekian H, Hoak RD, Kabler PW, Katz M, Keefer CE, Manganelli R, McGauhey PH, Miles HJ, Mohlman FW, Moore WA, Oxford HE, Rohlich GA, Ruchhoft CC, Sanborn NH, Setter LR, Trebler HA, Van Kleeck LW (1952) 1951 Literature review—a critical review of the literature of 1951 on sewage, waste treatment, and water pollution. Federation of Sewage and Industrial Wastes Associations. Sewage and Industrial Wastes 24(5):541–641

Snell JR, Corrough HM (1970) Land disposal of hazardous chemicals. In: Howe RHL (ed) Hazardous chemicals handling and disposal. The proceedings of the first symposium on hazardous chemicals handling and disposal. The Institute of Advanced Sanitation Research

Sperr FW (1921) Disposal of waste from gas plants—Report of 1921 Committee

Stanley WE, Eliasson R (1960) Status of knowledge of ground water contaminants. Department of Civil and Sanitary Engineering, Massachusetts Institute of Technology, Cambridge. December

State of New York (1900) Section 52, Laws of the State of New York passed at the 123rd Session of the Legislature. January 3–April 6

State of New York (1912) Laws of the State of New York passed at the 135th Session of the Legislature. January 3–March 29

State of New York (1913) Chapter 508, Section 247, Laws of the State of New York passed at the 136th Session of the Legislature. January 1–May 3

Stone R (1953) Land disposal of sewage and industrial wastes. In: Wisely WH, Orland HP, Mohlman FW (eds) Sewage and industrial wastes, vol 25, Jan-Dec. Federation of Sewage and Industrial Wastes Associations, Champaign, IL, pp 406–418

U.S. Congress (1976) Toxic substances control Act. 15 USC Section 2601

U.S. Congress (1979) Hazardous waste disposal: report together with additional and separate views. House of Representatives Ninety-Sixth Congress, Committee on Interstate and Foreign Commerce, Subcommittee on Oversight and Investigation Rep., U.S. Government Printing Office, Washington, DC

U.S. Department of the Interior (USDOI) (1968) Water quality criteria. Federal Water Pollution Control Administration, National Technical Advisory Committee to the Secretary of the Interior, Washington, DC

U.S. Engineer Office (1925) Report on investigation of pollution of navigable waterways and their tributaries. First District, New York City—Southern Section, New York

USEPA (1977) The report to congress: waste disposal practices and their effects on ground water. January

USEPA (1996) Catalyst for environmental responsibility. EPA 520-F-96-003, Narragansett, RI. Spring

USEPA (2001) Land disposal restrictions: summary of requirements. Offices of Solid Waste and Emergency Response and Enforcement and Compliance Assurance. EPA530-R-01-007. August

USPHS (1913) Municipal ordinances, rules, and regulations pertaining to public health. Compiled by Direction of the Surgeon General, JW Trask, Reprint No. 121 from Public Health Reports, January 26, 1912, to October 4, 1912, Inclusive. Government Printing Office, Washington, DC

USPHS (1915) Municipal ordinances, rules, and regulations pertaining to public health. Public Health Reports, January 9, 1914, to October 2, 1914, Inclusive. Government Printing Office, Washington, DC

USPHS (1917) Municipal ordinances, rules, and regulations pertaining to public health. Reprint No. 364 from Public Health Reports, 1915–1916. Government Printing Office, Washington, DC

Willien LJ (1920) Disposal of wastes from gas plants—report of 1920 Committee

Wilson DG (1977) History of solid-waste management. In: Wilson DG (ed) Handbook of solid waste management. Nostrand Reinhold, New York

Chapter 3
Environmental Sampling

Abstract Effective environmental sampling has two parts—representative sampling and accurate measurements. Both offer challenges. Performing successful sampling requires consideration of sufficient sample number, appropriate locations, and representative sample type (e.g., shallow, deep, composite). Ultimately, we attempt to represent large environmental volumes with a few small samples. Thus, careful thought is required to ensure representative sampling, and several "textbook" strategies exist to accomplish this.

Keywords Sampling · Representative · Data quality objectives · Composite samples · 95UCL · Monitoring

Introduction

Environmental sampling is like sausage—appealing, but it is unclear what is in it. Today, we sample or monitor many environmental elements—air quality, drinking water, sewage, landfills, hazardous waste sites, indoor air, paint, and the list goes on. There are probably thousands of environmental samples collected every minute.

Environmental measurement has two important parts—collection of representative samples and analysis. Many excellent references exist to instruct how to do both, but it is very difficult given actual field conditions, so a third step, data interpretation, is also paramount (Gilbert 1987; USEPA 1980, 1988). This chapter describes the kinds of issues raised by environmental sampling so as to provide a better understanding about what the results really mean.

Representative Sampling

Where and how to sample is the challenge. As described in Gilbert (1987), there are several statistically based sampling strategies, such as grid, transect, or random. That text and many US Environmental Protection Agency (USEPA) documents also

N. Shifrin, *Environmental Perspectives,* SpringerBriefs in Environmental Science,   19
DOI 10.1007/978-3-319-06278-5_3, © The Author(s) 2014

provide guidance on how many samples are required for various problems or objectives. Quality control of both the sampling and the laboratory analysis is also critical.

Successfully representative sampling begins with clear objectives, but it is amazing how often this fails. For example, most hazardous waste site sampling is biased to sample the dirty parts (e.g., near the leaking storage tanks). But the dirtiest parts (e.g., where the pure chemicals are) are often not sampled because: (1) it is obvious those areas are very dirty and (2) sending such dirty samples to the laboratory can create analytical problems. In addition, obviously cleaner areas often are sampled less. Knowing this helps with the interpretation of the study. For example, if the data are used for a human health risk assessment, which should assess the potential dangers from realistic exposures (e.g., often random encounters in time and space) to the hazardous waste site, knowing that the sampling was biased should at least lead to the interpretation that the assessment will overestimate the risks.

Almost every environmental sample has issues of representativeness. For example:

- Is an air sample in the right downwind direction from the source (knowing that the wind shifts)?
- Is an indoor air sample also collecting confounding chemicals from the cleaning solutions under the sink?
- Is the drinking water sample collected too soon after turning on the tap, so the lead in the pipe solder is distorting the result?
- From what depth should the river sample be taken?
- From what depth should the sediment sample be taken?
- What soil intervals in the 100-foot-deep boring should be sampled?
- How frequently should the groundwater well be sampled (knowing there are seasonal differences in groundwater)?
- How should the pit be sampled (e.g., sides, bottom)?
- Should we analyze the whole water or a filtered sample?

The answer to all of these questions is: It depends on the objective. The best environmental sampling programs have a clear understanding of objectives with correspondingly appropriate sampling design and careful sampling technique. That is why USEPA has gone to great lengths to develop guidelines and regulatory requirements for data quality objectives (USEPA 2000) and sampling plans (USEPA 2002a, b).

Samples can be individual or composite. Composite samples are a mixture of several samples from an area. They offer the benefit of understanding an average condition for less cost, but have the disadvantage of allowing no understanding of variability. Compositing techniques, such as the number of samples and sample mixing, can affect the result. The use of compositing can often be defined by the sampling objective. For example, a composite sample of a pile of waste may be sufficient when deciding where that waste might be disposed, but individual samples of a property may be required if the future use of portions of that property needs to be determined.

Sampling is an attempt to understand what truly exists without measuring the whole thing. That is the notion behind the 95% upper confidence limit (UCL) on the mean (95UCL). The mean of a group of samples is often assumed to represent some fundamental characteristic (e.g., how dirty, in general?). Because it is usually not possible to measure the "whole thing," some practical number of samples is taken as being representative, and an "error bar" (e.g., the 95UCL) is superimposed to ensure statistically that the true mean lies somewhere below it. Thus, with 95% confidence, the maximum of the true mean can be identified. The 95UCL is a function of the sample size and the range of the data—the fewer the samples, the less confidence there is that the data represent the true mean, so a higher 95UCL is needed, particularly for data with a large range.

The range, or spread of the data distribution, gives insight into how well a mean may represent a characteristic of interest. This "spread" is often represented statistically by the standard deviation or variance. The 95th percentile is the value below which 95% of the data points exist. It is different from the 95UCL. These statistical measures are commonly used to understand the nature of a data set, but are worthless if the samples are not representative of the sampling objective.

It is also useful to understand the difference between sampling and monitoring, because the two have different objectives. The objective of sampling is most often to understand the quality of some environmental compartment—sediment quality, for example. The objective of monitoring is usually to measure performance. One of the largest wastes of money in environmental measurement is the use of a sampling network left over from the characterization of a hazardous waste site to monitor the remedy. It is often done, because it is there, but it is often not necessary to use such an extensive network. Again, the design of a representative monitoring program should begin with a clear definition of the objective.

Compared to 50 years ago, environmental sampling has become sophisticated and common. But its reporting remains critical to understanding the data. There are good and bad reports, just as there is good and bad sampling. Environmental reports today still often emphasize a litany of numbers. Sometimes those numbers are compared to benchmarks—for example, to drinking water standards or background concentrations—which adds some perspective. (But remember, if the data are not representative, the benchmark comparisons can be misleading.) Sometimes the benchmarks themselves are misleading. For example, a geometric mean of background samples (i.e., not affected by the "source") is often used to represent the background benchmark, but given that the most extreme point in the background data set is still background, assuming the sampling was representative, why is that maximum not considered to be background? Sometimes it should be, but it depends on the distribution of the two data sets.

One way to judge a good environmental sampling report is to consider if it provided true understanding of the issue. Do not settle for less. The difference between data gaps and poor presentation of the data should be distinguished. Environmental sampling can legitimately be iterative, because there are true surprises in the field, and earlier rounds of data sometimes help inform about data gaps. But often, more sampling is used as a substitute for poor understanding of the data already collected.

Accurate Measurement

If truly representative samples have been collected, they still need to be measured appropriately and accurately. For example, if a chlorinated solvent is the concern, the measurement of trace metals does not seem appropriate. If the concern about that chlorinated solvent is the drinking water maximum contaminant level (MCL), a high-detection-limit method is not accurate enough. *Standard Methods* (APHA et al. 2012) is an analytical methods textbook that has been published over 20 times since 1905 as methods and concerns have evolved. Agencies have issued regulatory-approved methods for soils (USEPA 1980), air (USEPA 1994), and water (USEPA 1997; SAEPA 2007) that have been perfected over time and include many quality assurances (Webb and McCall 1973). There are multiple methods for target analytes, depending on the objective, so appropriate method selection can be important.

Some environmental measurement targets are actually groups of compounds, which require sophisticated analysis and interpretation. For example, polychlorinated biphenyls (PCBs) are a group of 209 compounds ("congeners") with varying numbers of chlorine atoms and positions around two attached benzene rings ("biphenyl"). Accurate measurement of PCBs is complicated by the fact that they were marketed and historically measured as Aroclors, which are determined by the percentage of chlorine (e.g., Aroclor 1242 has 42% chlorine by weight), not by a precise grouping of congeners. Webb and McCall (1973) developed an analytical method for identifying and quantifying these Aroclors.

In addition, Aroclors and dichlorodiphenyltrichloroethane (DDT) can be confused on the gas chromatograph, so appropriate preparation of samples must be used to minimize this (Armour and Burke 1970). Adding to the difficulty is today's trend to analyze PCB congeners rather than Aroclors, so comparison of modern data to historical data can be difficult. Although congener analysis can help avoid PCB "weathering" misinterpretation, modern congener analysis often does not include all 209 congeners. Incomplete data may preclude an often-used parameter called "total PCBs" or further hinder comparison with historical data.

More analytical/interpretation issues arise with other environmentally pertinent groups like dioxins and polycyclic aromatic hydrocarbons (PAHs). In addition, speciation can be important. For example, lead sulfide is considered less toxic than lead acetate (USEPA 2007). Similarly, hydrogen cyanide is very toxic, but ferric ferrocyanide, commonly associated with former manufactured gas plants, is nontoxic (Shifrin et al. 1996)—so much so that it is used in children's crayons. Many environmental studies simply report "cyanide," which might make regulators shudder, but that cyanide might be present as relatively inert species. Speciation is rarely performed in environmental sample analysis.

Precision is the ability to reproduce the result; it is different from accuracy, which is the ability to be right. In the laboratory, precision is often ensured by analyzing duplicates. The accuracy problem with duplicates is, if the results are disparate, which one is right? Triplicates might help solve this problem if two of the disparate results are closer together. Triplicates are rarely analyzed, however. Another issue to consider about duplicates is whether a sample was split in the field or in the laboratory. Field splits might also be two separate samples or one sample split in two. Guidance exists to determine which kind of duplicate is appropriate, but it is also a matter of the objective of the duplicate. For example, is the duplicate a measure of analytical or sampling precision?

Amazing Results

Given all the nuances involved with environmental sampling, it is amazing that the environment can be characterized at all. But it can. Environmental professionals are trained to measure the environment and interpret the results. It is important to recognize the limitations in environmental sampling and to use competent professionals who can minimize those limitations and explain them. Environmental data collection is expensive, but there are enough issues involved with it to avoid simply choosing the least-expensive option.

References

American Public Health Association (APHA), American Water Works Association (AWWA), Water Environment Federation (WEF) (2012) Standard methods for the examination of water and wastewater. 22nd edn

Armour J, Burke J (1970) Method for separating polychlorinated biphenyls from DDT and its analogs. J AOAC 53(4):761–768

Gilbert R (1987) Statistical methods for environmental pollution monitoring. Wiley, New York

Shifrin N, Beck BD, Gauthier TD, Chapnick SD, Goodman G (1996) Chemistry, toxicology, and human health risks of cyanide at former manufactured gas plant sites. Regul Toxicol Pharmacol 23:106–116

South Australia Environmental Protection Authority (SAEPA) (2007) Regulatory monitoring and testing, groundwater. Adelaide

USEPA (1980) Test methods for evaluating solid waste, SW 846. Office of Water and Waste Management, Waste Characterization Branch, Office of Solid Waste, U.S. Environmental Protection Agency, Washington, DC, May. (updated thereafter)

USEPA (1988) Guidance for conducting remedial investigations and feasibility studies under CERCLA. OSWER Directive No. 9355.3-01. October

USEPA (1994) Quality assurance handbook for air pollution measurement systems. Office of Research and Development. EPA/600/R-94/038a

USEPA (1997) Guidelines for preparation of the comprehensive state water quality assessments (305(b)) reports. Office of Water. EPA-841-B-97-002A

USEPA (2000) Data quality objectives process for hazardous waste site investigations. Office of Environmental Information. EPA/600/R-00/007

USEPA (2002a) Guidance for quality assurance project plans. Office of Environmental Information. EPA/240/R-02/009

USEPA (2002b) Sampling and analysis plan (SAP) guidance and template. R9QA/001.1

USEPA (2007) Estimation of relative bioavailability of lead in soil and soil-like material using in vivo and in vitro methods. OSWER9285.7-77

Webb R, McCall A (1973) Quantitative PCB standards for electron capture gas chromatography. J Chromatogr Sci 11(7):366–373

Chapter 4
Environmental Analytical Chemistry 101

Abstract Environmental chemical analysis has two parts—sample preparation and analysis. Today's instruments measure extractions of the chemical that existed in an environmental medium, such as soils, so the first challenge is to remove the target chemical ("analyte") purely from the medium. This step often is complicated by the extraction of unwanted chemicals, causing interferences, or by difficulties in removing all the analyte from the medium. Sometimes, the analyte is actually a group of chemicals, such as polychlorinated biphenyls (PCBs), which offer additional challenges. Common analytical instruments include gas chromatographs, mass spectrometers, and atomic absorption or inductively coupled plasma spectrophotometers. Operating each requires skill with sample preparation and the instrument, along with a chemist's interpretation of the electronic output.

Keywords Precision · Accuracy · Matrix · Gas chromatography · Detection limit · Analytical methods

Introduction

Laboratory analysis of environmental samples has evolved from the limited "wet chemistry" methods of the pre-1970s to today's powerful and sensitive instrument methods. Guiding this sophisticated technology are a host of unique regulatory and trade group methods (USEPA 1986; APHA et al. 2006; ASTM International 2013; AOAC International 2013), resulting in multiple ways to analyze any particular chemical. The accuracy, precision, and detection limits that can be achieved vary according to the method used.

Environmental contaminant analysis has two basic parts—sample preparation and analysis. Some requirements are prescriptive (e.g., "Do this after that"), while others are performance based (e.g., "Do it any way you want, but you must meet these requirements"). Each part of the analysis affects what is possible for accuracy, precision, and detection levels. Unfortunately for a data user, a true value of "100" might easily be reported as "50" or "200." Similarly, a result that fails a bright regulatory line of "5" by a laboratory reporting of, say, "5.1" could easily have a true value considerably lower than the regulatory limit.

N. Shifrin, *Environmental Perspectives*, SpringerBriefs in Environmental Science, DOI 10.1007/978-3-319-06278-5_4, © The Author(s) 2014

This chapter describes the fundamentals of laboratory analysis of environmental samples.

Fundamentals

Environmental analytical chemistry is filled with simple concepts that have subtle complications and confusing jargon. For example, a "detection limit" has a name that is often misused or misunderstood. Worse, some environmental measurements are inherently vague and require expert interpretation subject to debate. A speedometer measurement of 50 mph is clear, but a "petroleum hydrocarbon" value of 50 mg/kg could mean different things to different experts—and nothing to nonexperts.

Cutting through the jargon to help clarify a general understanding, some fundamental concepts are described below.

A matrix is the material containing the contaminant. Soil, water, wastewater, and biological tissue (e.g., fish) are common matrices. The matrix affects the sample preparation technique and can affect the analytical results because some chemicals ("target analytes") are more difficult to remove from some matrices. The first step of environmental analytical chemistry is to remove the target analyte from the matrix, so it can then be measured by an instrument. For example, pesticides are extracted from a water sample by passing the sample through a column filled with specially reactive beads. The pesticides, which preferentially adsorb (stick) to the beads, are then removed from the beads with a small amount of solvent, resulting in a pesticide concentrate for instrument analysis. A pesticide that adsorbs poorly to the beads would result in poor recovery, meaning that the actual (e.g.) 100 μg/L present might only result in 25 μg/L measured at the instrument (i.e., have a 25 % recovery).

Analytical methods specified by various state and federal programs have different requirements for recoveries of target analytes. To some degree, the effect of recoveries on results could be minimized by dividing the result by the recovery (methods include quality control requirements to keep track of recoveries), but this is rarely done if the recovery rates are within the acceptable ranges defined by the regulatory programs.

Chemical loss during sample preparation is another way analytical accuracy can be affected. Sample storage, although not precisely a preparation step, can also affect accuracy if the target analyte degrades, reacts, or volatilizes prior to analysis. Method-stipulated storage time and temperature requirements help minimize such losses, such as for analytes like volatile organics.

Precision and accuracy are different, and both are critical concepts. Precision is the ability to reproduce the same result (i.e., "100" is measured as "100" every time). Different methods define specific precision requirements, which are examined by repeated measurements of the samples and laboratory standards. Matrix standards examine replication of the entire analysis, whereas instrument standards examine the precision of the instrument portion of analysis, described below. Standards are also used to measure accuracy, which is the ability to report the true value. Internal standards and surrogates are chemicals with properties similar to environmental contaminants, but are unlikely to be found in environmental matrices, which are added to track recoveries. If a standard precisely prepared to be "100" yields a result of "100," the analysis is accurate.

Perhaps the most misunderstood element of analytical chemistry is the detection limit. The terminology varies by program, as does how detection limits are derived, but the basic concept from lower to higher is: (1) What can be sensed, (2) what can be quantified, and (3) what is the minimum requirement for quantitation prescribed by the method? For example, using a sound analogy, nothing can be heard below background noise, but a sound just above background noise can be sensed but probably not understood. At some higher sound level, it might be understood as a word, but the word might be poorly distinguished. At some slightly higher level, the word can be understood as the same word every time it is heard.

In analytical chemistry, there is an "instrument detection limit," which is the lowest signal measureable by an instrument above background noise (electronic or chemical); a "method detection limit," which is the lowest concentration above zero that can be detected as present in a given sample with 99% confidence; and a "practical quantitation limit" (sometimes given names like the "contract required quantitation limit" or "limit of quantitation"), which is the lowest reliable numerical concentration that can be routinely reported. Note the distinction between presence and concentration.

Two other fundamental concepts to help understand environmental analytical chemistry are analyte groups and methods. Common environmental contaminants are consolidated into analyte groups according to their properties. Volatile organic compounds (VOCs), semi-volatile organic compounds (SVOCs), trace metals, pesticides and polychlorinated biphenyls (PCBs), and dioxins/furans are some of the common analyte groups. The actual constituents within those groups vary by method. Thus, the Superfund VOC list is different from the Clean Water Act (CWA) VOC list. Laboratory services are typically purchased according to such groups. For example, a drinking water sample analyzed for trichloroethylene (TCE) is typically purchased as part of a VOC analysis that includes more than 20 other chemicals under the Safe Drinking Water VOC list. Metals are also bundled into groups depending on the applicable regulatory program (i.e., Resource Conservation and Recovery Act, RCRA, 8 metals; CWA priority pollutant 13 metals; or Superfund target analyte 23 metals). Petroleum hydrocarbons are a less straightforward analyte group. Some methods report "total petroleum hydrocarbons"; other methods offer a finer separation of petroleum hydrocarbon components like "gasoline-range organics" and "diesel-range organics," or carbon ranges representing petroleum fractions (e.g., 5–12 carbons for gasoline, 9–15 carbons for kerosene, and 9–24 carbons for fuel oil). Some chemicals cross group boundaries, such as naphthalene, which is analyzed in both VOC and SVOC groups, and the results typically differ depending on the method.

Today, analytical methods are identified by unique method numbers. The method number tells much about the technique used, detection limits, precision, and instrument, in addition to the regulatory program requiring the analysis. For example, the Safe Drinking Water Act requires VOC samples to be analyzed using US Environmental Protection Agency (USEPA)'s 500 series methods, the CWA requires USEPA's 600 series or Standard Method's 6200 series, and for RCRA and Superfund they are generally run under USEPA's 8000 series. Typically, analytical services are purchased by defining the method (e.g., USEPA 8260, which means VOCs by gas chromatography, GC, coupled with mass spectrometry, MS).

The Instruments

The workhorse of today's analytical laboratory is the gas chromatograph (GC) for organic chemicals. Its function is simple—separation of multiple target analytes so each one can be identified and quantified individually. It does this by transporting a gas laden with the target analytes through very long, inside-coated tubes ("columns" or "capillaries") where each analyte is retarded on the tube to different degrees, depending on its chemical properties. Each chemical exits the tube after a precise "retention time" roughly predictable by its solubility relative to the other chemicals—more soluble chemicals exit first, less soluble last.

Each chemical must be measured as it exits. Each analyte is identified by comparing its retention time (through the GC tube) and concentration to runs using prepared laboratory standards. Analyte concentrations are measured by a variety of detectors attached to the GC. Typical detectors include the flame ionization detector (FID), which measures the ions formed as the gas leaving the GC is burned in a hydrogen flame; the electron capture detector (ECD), which measures the electrons formed after the gas is bombarded with radioactive particles; and the mass spectrometer, which essentially smashes the molecules of an existing chemical into distinct mass-to-charge entities.

Trace metals are typically measured by flame or furnace atomic absorption spectrometry (AAS) or inductively coupled plasma (ICP). As astronomers watching stars know, each element absorbs specific wavelengths of light by shifting electrons to different orbitals when "excited" and emits light of specific wavelengths as the orbitals "decay." After calibration with standards of specific concentrations, AAS machines can be used to identify and quantify the concentrations of about 70 elements in a combustion flame. Special techniques, like the "graphite furnace" method, can be used to measure very low levels of environmental contaminants like mercury. ICP methods use extremely high heat to "excite" atoms and ions and measure the characteristic electromagnetic radiation by ICP-atomic emission spectroscopy (ICP-AES) or ICP-MS. ICP can analyze samples for multiple trace metals at the same time.

Today's modern laboratory has numerous other instruments to measure very low levels of exotic and routine chemicals, but the instruments described above are typically what are used for analysis of many environmental samples.

Analysis

Many steps are involved in taking a jar of dirt or water and reporting, for example, a dioxin concentration of one part per quadrillion (1 picogram per liter). After a sample is collected and transported to the laboratory, the target analytes are extracted (differently for each analyte groups). If necessary, interfering compounds are removed ("cleanup") and the extraction conditioned for instrument injection (e.g., diluted in its carrier solvent if its concentration is so high it would otherwise overwhelm the instrument). Quality control standards (i.e., "matrix spikes" and "surrogates") might be added. The prepared solution is injected into an instrument (e.g., GC or AAS), and the detector signal is recorded and converted to a concentration.

How to Consider the Results

It may now be clear that the results depend on the method, laboratory, and analyst, and that "the number" may not be exact. Although these observations mean that some regulatory exceedances or low-level detections might be debatable, it does not mean that environmental analysis is hopeless. It simply means that understanding the analyses can lead to enlightening insights into the data. It is critical to look carefully at the laboratory report to ensure that appropriate recovery and detection limits have been achieved, and that quality control requirements have been met. In most cases, even poor data (e.g., data with exceptionally low recoveries) can be used, but they must be viewed in context. This is especially important for chemicals that have detection limits near regulatory standards, when slight exceedances may result in the need for significant remedial actions.

Results can differ from time to time or place to place for many natural reasons. Knowing the analytical limitations of environmental data might help understand whether variability is actual or an artifact of the analysis.

References

American Public Health Association (APHA), American Water Works Association (AWWA), Water Environment Federation (WEF) (2006) Standard methods for the examination of water and wastewater, 22nd and online edition. American Public Health Association, Washington, DC. http://www.standardmethods.org. Accessed 17 July 2013

AOAC International (2013) Official methods of analysis of the association of official analytical chemists. Arlington. http://www.aoac.org. Accessed 17 July 2013

ASTM International (2013) ASTM Standards and Engineering Digital Library (SEDL). West Conshohocken. http://www.astm.org/DIGITAL_LIBRARY/index.shtml. Accessed 17 July 2013

USEPA (1986) Test methods for evaluating solid waste, physical/chemical methods. EPA Publication SW-846

Chapter 5
Environmental Forensics

Abstract Environmental forensics consists of both advanced evaluation of typical chemical analyses and advanced chemical analyses. Forensic analysis is often used to differentiate between sources or to examine the timing of historical releases. Some techniques include chemical fingerprinting, speciation, radionuclide dating, microscopic analysis, and statistical analysis. Although it has limitations, forensic analysis can be useful to extract much more from the data than usual.

Keywords Environmental forensics · Hydrocarbon fingerprinting · Isotopes · Chemical species · Differentiation · Timing · Cause and effect

Introduction

Forensics has many roles in the law, from murder trials to accounting. It is also applied to environmental issues and can be useful in environmental litigation. Environmental forensics has its own journal (International Society of Environmental Forensics', ISEF's, *Environmental Forensics*), which spans research ranging from factory evolution and its historical waste discharges to source differentiation by chemical data interpretation. This chapter focuses on the latter. As described in the *Forensic Chemistry Handbook* (Kobilinsky 2012), environmental issues are only one topic of this science. Forensic chemistry also involves the studies of explosions, fires, paint, ink, and human samples (e.g., tissues, drugs, DNA).

Laboratory analysis of environmental samples typically reports concentrations of chemicals organized by standardized groupings, such as volatile organic compounds (VOCs), trace metals, and semi-volatile organic compounds (SVOCs). The basic data reports for these analyses are sometimes sufficient. For example, if a simple groundwater plume contaminated with trichloroethylene (TCE) needs "chasing," measuring its concentration trend along a downgradient transect is a simple matter of taking successive groundwater samples and analyzing for that chemical. But if TCE's natural attenuation is also an issue, analysis of its degradation "daughters" and other degradation clues, such as redox potential, becomes important, and a forensic approach in its simplest form is more appropriate.

N. Shifrin, *Environmental Perspectives,* SpringerBriefs in Environmental Science, DOI 10.1007/978-3-319-06278-5_5, © The Author(s) 2014

These advanced forms of data generation and interpretation can sometimes clarify the mysteries of a source and the fate of chemicals in the environment. Forensic analysis can be a powerful tool, but it also has limitations.

Typical Applications and Advantages

Forensic analysis can be used to differentiate sources, such as the type of hydrocarbon (e.g., tar or oil), polychlorinated biphenyl (PCB) releases (e.g., paper mills or electrical equipment), dioxins (e.g., a chemical plant or fire), and lead in soils (e.g., from mining, smelting, or paint). It can also be used to date historical releases, such as differentiating tars based on known dates of manufacturing processes. In addition, it can be used to understand fate and transport processes, such as the example of natural attenuation of chlorinated solvent noted above.

In its simplest form, forensic analysis looks deeper into the data already collected. Sometimes this is called "fingerprinting," because some types of contamination have recognizable chromatographs (the "raw" data plot from gas chromatography, GC). For example, examination of a gas chromatograph can quickly show if hydrocarbon contamination is petrogenic (petroleum derived) or pyrogenic (fire derived). Similarly, chromatographs sometimes can reveal information about PCB "weathering" (partial dechlorination of some of the 209 PCB congeners), which can offer insight into both a source and the timing of its release. Double-ratio plots of polycyclic aromatic hydrocarbons (PAHs), which are typically measured in a routine semi-volatiles analysis, can reveal tar types (e.g., coal tar or carbureted water gas tar) or whether the source is tar at all.

Most often, "total" analyses are performed, such as total lead or total cyanide. A deeper look into speciation, however, can reveal more about a source of contamination. For example, knowing if the lead is lead sulfide or lead oxide can differentiate between a mining and a paint source. Ferric ferrocyanide is mostly unique to old manufactured gas plants (MGP), so knowing that species exists tells much about the source. Unfortunately, even forensic analysis is limited to a deduction from the available analytical methods, which do not measure directly that form of cyanide (Shifrin et al. 1996).

Measurement of contamination in successive depths of a sediment bed can offer insight into release timing, especially if coincident isotopic (e.g., cesium 137) or even pollen data are available. This dating example suggests that forensic analysis can be enhanced with coincident data not typically measured in environmental studies. Proper planning is thus required. Enhanced data requirements for dating often include an additional analyte that has a unique time characteristic, such as the appearance and then disappearance of cesium 137 due to atmospheric nuclear testing. In other cases, coincident data may be useful because they are associated with particular sources. For example, biomarkers persist in crude petroleum as the complex molecules that existed in the original plant cells that ultimately became the petroleum. Measurement of those biomarkers in petroleum today can sometimes reveal the geographic source of a hydrocarbon.

Microscopic analysis can also be a useful adjunct to chemical analysis. For example, such an examination of a subsurface waste zone might reveal certain particles like coal, ash, or fibers that can help differentiate sources.

Statistical analysis of a data set is another forensic tool. For example, polytopic vector analysis (PVA) is a pattern recognition method applied to data to differentiate sources such as dioxins and PAHs. PAHs are one of the most intriguing challenges for forensic analysis because there are so many different sources, and the group of chemicals is ubiquitous. A typical environmental analysis looks for 16 of them.

PAHs are a group of compounds, each having a different arrangement of fused benzene rings and with varying properties and toxicities (benzo(a)pyrene is often considered the most toxic). PAHs are generally formed from the incomplete combustion of organic materials—such as from vehicles, manufacturing (e.g., former MGPs), tobacco smoke, volcanoes, backyard barbeques, forest fires, and incineration—but petroleum is also a non-combustion source. Soils generally have an elevated "background" level of PAHs from both anthropogenic and natural sources (Bradley et al. 1994). The PAH pattern can sometimes help determine a source (e.g., with double-ratio plots or PVA). The type of PAH is sometimes revealing, such as a methylated PAH, which indicates a petrogenic source. Often, however, PAHs are so mixed and ubiquitous from so many sources that even a forensic analysis offers limited insight.

Limitations

The above-mentioned example of PAHs demonstrates that forensic analysis is useful but not omnipotent. The environment and its contamination pose limitations such as the PAH example; other issues are posed by insufficient data or analytical measurement limitations. Expertise is required to both design a proper study that will support a forensic analysis and actually perform the analysis.

Environmental limitations are often created when too many sources are mixed. The soup may just be too thick to see through. For example, a mixture of several chlorinated solvents, even if from separate sources, might not be possible to differentiate. In other cases, weathering effects might make the original source unrecognizable. For example, mine-derived waste may become unrecognizable after being oxidized by blowing downwind and sitting on the aerated ground surface. In still other cases, the concern may have too many causes. For example, PCBs in fish might be the concern in a river with many different sources of PCBs. A spatial analysis, which is not really "forensic analysis," may help, but in the PCB example, the fish swim around, which may helplessly complicate the analysis.

Insufficient data are often a case of poor planning. Environmental studies should be designed first by knowing if forensic analyses will be used. A not-uncommon example complaint is: "If only we had collected the cesium-137 data, we would know if that xyz disappearance at 3 ft in the sediments corresponds to the plant startup date." A supplemental study may provide the needed information, but often there will be lingering doubts. For example, a return to the sediment core location of the 3-foot disappearance to take new cesium data can likely time the disappearance, but how important will it be that it is not exactly the same location?

Budget considerations often add a tension to ensuring sufficient data for forensic analysis. Environmental sampling is expensive, and the additional data required for some forensic analyses make it even more expensive. Whether the additional cost is worth it will depend on the likelihood that the additional laboratory analyses can serve their purpose and how strong the forensic analysis will be to prove its point.

The lack of appropriate analytical methods also can limit forensic analysis. The cyanide comment above is an example, since cyanide speciation can be quite difficult. Sometimes methods are newly improved and become more appropriate, but the forensic analysis may need to compare new data to old, which still used inadequate methods. For example, recently developed PCB congener analysis offers powerful data to help differentiate sources and understand weathering, but this often requires a comparison to older Aroclor data, which are not nearly as powerful.

When to Use Environmental Forensics

Among other uses, and recognizing its limitations, environmental forensics may be used to:

- Differentiate among sources
- Reject or identify a suspected source
- Time a source
- Examine fate and transport
- Understand cause–effect relationships

Expertise is required to consider when environmental forensics will be useful and what approaches to use. Highly specialized expertise, such as a specialized laboratory, may be required, but big-picture expertise is also useful to integrate pieces and explain what they really mean. The challenge of when to use environmental forensics is to know what kinds of analyses are applicable, how likely the results will be useful, and whether the added expense will be worth it.

References

Bradley LJN, MaGee BH, Allen SL (1994) Background levels of Polycyclic Aromatic Hydrocarbons (PAH) and selected metals in New England urban soils. J Soil Contam 3(4):3493–3561
Kobilinsky L (ed) (2012) Forensic chemistry handbook. Wiley, Hoboken, 504 p
Shifrin NS, Beck BD, Gauthier TD, Chapnick SD, Goodman G (1996) Chemistry, toxicology, and human health risk of cyanide compounds in soils at former manufactured gas plant sites. Regul Toxicol Pharmacol 23:1061–1016

Chapter 6
Environmental Data Visualization

Abstract The graphical display of environmental data can help interpret their meaning. For example, a three-dimensional display of measurements in an environmental space shows a picture that would otherwise need to be imagined, and often incompletely, in the mind of someone reviewing otherwise tabular data. Fortunately, many tools aid in data visualization, such as database programs, GIS, and modeling, all of which are accessible through inexpensive personal computer software. Future developments may offer exciting improvements.

Keywords Three-dimensional · 3D · Data posting · Metadata · GIS · Database · Bar chart · Isoconcentration contours

Introduction

Environmental data can be numerous and onerous. Graphical presentation of the data usually makes review simpler and offers insights that would otherwise be difficult to see in numerical tabulations. Effective data visualization is part science and part art. Fortunately, powerful computer tools now exist to enable wonderful data presentations, limited primarily by the creativity of the presenter who must also understand possible limitations of the data to avoid misrepresentation. This chapter presents some simple data visualization examples.

A Picture Is Worth Many Tables

Spreadsheet and database software make it simple to create seemingly endless tables of data capable of causing anyone's eyes to glaze over. A picture (e.g., a graphic of some type) is far superior, but the challenge is to understand what to show or emphasize. Understanding emphasis is science. For example, showing only the data above drinking water standards, the mean along with variability, or just the most frequently encountered contaminant are points of emphasis that might be made depending on an understanding of what is important. Presenting it is art. It is not art as in entertainment, but it is art in the form of capturing the eye and conveying understanding.

N. Shifrin, *Environmental Perspectives*, SpringerBriefs in Environmental Science,
DOI 10.1007/978-3-319-06278-5_6, © The Author(s) 2014

Most people understand information better by picturing it than by reading words or tables. Thus, a picture of an environmental site (e.g., a photo or a map) with a proper scale and reference to the larger area is more effective than descriptive text like, "100 feet southwest along Main Street...." Subsequent information can be built upon this picture. For example, a comment that a sample was taken from the bottom of a riverbank will offer better understanding if an accompanying picture shows that the riverbank is steep and tall.

Data, Metadata, Concepts, and Tools

The notion of data is probably obvious. Metadata are the data about the data, such as the x, y, z coordinates of a sampling point; the analytical method used; and the detection limit. Metadata can also be similar data across numerous studies. A database stores the data and the metadata, but a database is more than just a software tool (of which there are many). A well-designed database efficiently stores the data and metadata in a relational way that allows for effective retrieval through queries. For example, a simple query of "all of the bedrock groundwater data" for a certain chemical will fail unless the database somehow "knows" what data are related to bedrock groundwater. Similarly, a query will give useless results if the data were not entered accurately.

Accuracy is facilitated today for environmental databases by electronic entry—electronic tables from the laboratory loaded directly to electronic tables in the database—with human error eliminated. Databases are often expensive to create, but they result in cost savings and better accuracy later, during data interpretation.

Today, databases often exist within Geographic Information Systems (GIS), a visually handy way to store data with metadata and to present related information visually. The various metadata can be overlain graphically as "layers." For example, a GIS might plot all the soil data along with a sewer map "layer" so that the viewer can consider if soil contamination has a relationship with leaky sewers. Some GIS systems have dozens of layers and become a visual model of an environmental site. The ability to "turn on" or "turn off" layers easily makes searching the data for correlations and presentation of related information more effective.

Another visual tool that has been greatly simplified over the years is data contouring. An "isometric" plot (sort of a misuse of the term, but commonly used) shows lines (two-dimensional, 2D) or contours (3D) of a constant quality, such as elevation or concentration, usually on a map of a site. A US Geological Survey (USGS) "quadrangle" mountain hiking map is an example. The classic isoconcentration map of 20 years ago has been greatly enhanced today with software that uses color shading to show the variations (2D) and color shading along with vertical contours for even better 3D visualization. The latter used to require hundreds of dollars and days of mainframe computer time-sharing cycles; today, it takes minutes on a PC with US\$ 500 software. One glance of such a 3D contour map can give

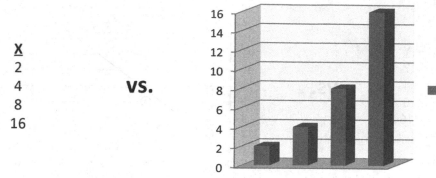

Fig. 6.1 Simple bar chart

a viewer instant understanding that might otherwise take hours (or an eternity) by viewing the same information in a table.

Common Data Visualizations

One of the simplest visualizations is the bar graph. Consider the enhanced "feel" from a bar graph presenting the same information as a list of numbers. A further improvement is a 3D bar graph (Fig. 6.1).

Slightly more complicated, but common, is a regression plot for a group of "x–y" data, where the line drawn is the "best fit" correlation of "x versus y"; and R^2, the regression coefficient, is a measure of how well the line fits the data (1 is perfect, below 0.9 is considered by many as a poor correlation; Fig. 6.2).

A powerful visualization is often shown with a 3D bar graph comparing a contaminant having multiple components (e.g., dioxin congeners) to standards for pattern recognition (Fig. 6.3).

As noted above, the classic 2D contour plot can be improved with 3D visualization techniques (Fig. 6.4).

A combination of tabular and visual information can be "posted" data on a map of a site. This technique can provide a tremendous amount of essential data (e.g., sample date, sample depth, key chemical concentrations) while visualized by location, which enables focusing without overwhelming. Data posting also saves time otherwise required to flip between a table and a map (Fig. 6.5).

Depth and time are important coincident dimensions for many environmental data, and 3D visualization can often incorporate these dimensions to enhance understanding.

These are just a few of the tools available to environmental professionals who must present and understand data. But they are just that—tools: A hammer and a saw still require a carpenter to build a house. It is important to understand that

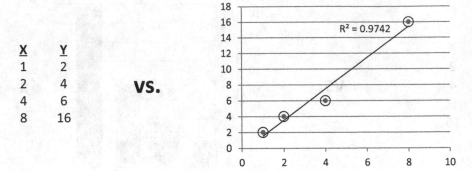

X	Y
1	2
2	4
4	6
8	16

Fig. 6.2 Scatter plot

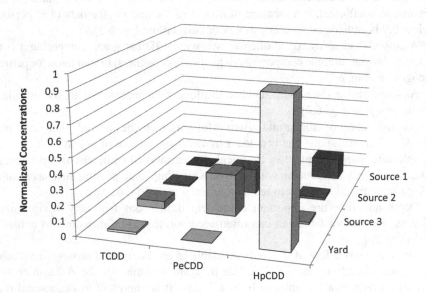

Fig. 6.3 Dioxin congeners

Fig. 6.4 Contour plots

TCE in Groundwater (mg/L)

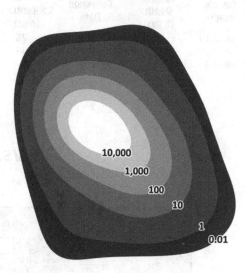

VS.

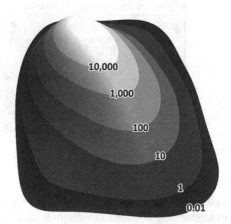

Sample Location	Sample Depth (ft bgs)	Collection Date	Mercury Concentration (mg/kg)
SB-101	10	06/26/2013	28
SB-102	12	06/25/2013	36
SB-103	10	06/26/2013	59

vs.

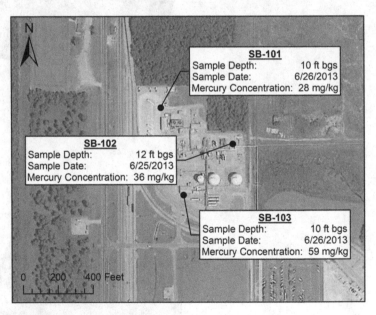

Fig. 6.5 Data post

these tools are much more than pretty pictures. They have become essential for understanding complex environmental issues. Skilled use of these tools will result in better solutions.

The Future

Since today's data visualization capabilities are so stunning over what existed 40 years ago, it is difficult to believe they could get much better. But they will. Some of the improvements will lie with progress in better 3D depiction of their 2D portrayals (e.g., holograms). Animations of varying 3D perspectives (still in 2D, but variable to give a better sense of 3D) exist today and will improve. For example, the view of a subsurface well field (and the sampling results) can be rotated in real time on a computer screen, including from subsurface perspectives, so that hidden areas from a single perspective and better perspectives of relative depths can be seen. The objective is to convey understanding, not a parade of knowledge. Environmental professionals with competent scientific expertise, creativity, and the right tools achieve this today, and it will get better in the future.

Chapter 7
Environmental Modeling

Abstract Environmental models are often used to describe the transport of contaminants when measurements are not possible—for example, in predicting future events. Models can be computer simulations or napkin-back equations. Using either requires an understanding of the simplifications made to the governing equations and to the many required input-, boundary-, and initial-condition parameter values. Model calibration involves setting realistic parameter values until the model output matches reasonably well with a set of measurement data. Model validation involves matching the model output to a second set of data using the calibrated parameter values. Packaged models exist for many environmental settings, including surface water, groundwater, river sediments, air, soil vapor, and indoor air.

Keywords Transport · Advection · Dispersion · Gaussian plume · Calibration · Verification · Simulation · MODFLOW · AEROMOD

Introduction

A *model* is the representation of a physical entity. Examples include model cars, architectural models of planned buildings, a wind tunnel, an equation representing a state of physics, or a computer simulation of an environmental process, such as groundwater contamination, sewage into a river, or climate change. Models are used to predict that which is difficult to measure, such as future events or complex cause–effect relationships. Environmental modeling might be "analytical" (an equation on a napkin) or a simulation, which often involves solving iterations of a series of related equations (most conveniently on a computer). Environmental modeling can help gain insight into processes (e.g., the important factors or cause–effect relationships), and it can be used to predict future events (e.g., what will the groundwater quality be if the source is removed?). There are many issues for creating good environmental models, and the adage "garbage in/garbage out" definitely applies. This chapter describes the fundamentals of environmental modeling, including when to model and when to avoid models, so that non-modelers can understand their utility and their frailties.

N. Shifrin, *Environmental Perspectives*, SpringerBriefs in Environmental Science, DOI 10.1007/978-3-319-06278-5_7, © The Author(s) 2014

Modeling 101

Environmental models are often used to describe the environmental transport of contaminants. Other types of environmental models exist, such as climate change models, but this chapter focuses on contaminant transport models. Such models have two parts: flow and transport. Transport often involves transformations along the way. Examples of transformations are chemical reactions, such as chromium reduction, or biologically mediated changes, such as the mercury methylation or the creation of acid mine drainage. The flow component of a model is most often a tremendous simplification of the universal Navier–Stokes 3D differential equations describing momentum and location of a fluid in space and time.

These greatly simplified versions of the Navier–Stokes equations then are adapted to computer code for simulation via finite elements of finite differences to approximate the continuous calculus by discrete "chunks" that can be manipulated by a computer. Many modern models can be thought of as a "Rubik's cube" grid, with each sub-cube having a set of equations that transfers mass and energy across each boundary to its neighbors according to certain rules.

Many models are referred to as "Gaussian plume" models, because they predict concentration trends that taper equally across a flow centerline. A graph of predicted concentration across the flow field and along the distance would look like a series of "bell-shaped" (i.e., Gaussian) curves. This is because the models use mathematically symmetrical "dispersion coefficients" to describe contaminant concentration reductions due to turbulence and diffusion with flow.

Models require input and boundary conditions. An example input condition might be the mass per second of a chemical emission from a smokestack. An example boundary condition might be "no flow" below a certain depth in a groundwater model "bottomed out" by bedrock. Often, these conditions are approximations. For example, the bedrock probably does allow some flow, but for practical purposes, zero flow can be assumed.

Some models have complex input requirements. For example, anisotropy is common for hydraulic conductivity in soils (i.e., different in the horizontal vs. vertical directions), and this parameter can also vary tremendously along lateral distances. Many times, only a handful of hydraulic conductivity data (sometimes none) exist for a site, but a groundwater model might still be attempted. Another example is fugitive emissions in air models. Air modeling of a smokestack is relatively straightforward, but if the same chemical of concern also blows off of piles, yard dirt, or out of factory windows, those "fugitive emissions" can play a critical role in the result, but they are very difficult to characterize as emissions. Many inputs also often vary with time.

This dark tunnel of parameter values is enlightened during model calibration. Calibration is the model preparation exercise where input parameters are varied and resultant model predictions are compared to a set of measurement data. If the model results compare reasonably well to the measurements (there are statistical tests, such as "residuals") and the input parameter values are within reason, the model is considered calibrated. Calibrated models should then be verified.

Model verification involves comparing the calibrated model to an entirely new set of measurement data. For example, a river model might be calibrated to a data set involving one streamflow, but how does it compare to a data set from a different streamflow? Successful verification adds tremendously to a model's certainty because, without it, there is more of a chance that the model calibration was simply a forced fit.

Although this discussion has emphasized computerized models, often called *simulations,* more simple "analytic" models sometimes suffice and can even be superior (see below). An analytic model is a simple equation describing a process or condition that can be solved "by hand." For example (not necessarily in reality), if air concentration on the other side of a window is always one half of that inside the room, $C_{out} = 0.5C_{in}$ is an analytic model of that condition and solvable by hand if C_{in} is known.

Basic Environmental Models

One of the first environmental models was offered by Streeter and Phelps (1925) to predict stream reactions to sewage discharges. Air modeling was consolidated by D. Bruce Turner in the 1960s (Turner 1961) resulting in his famous "Turner workbook" (Turner 1967), although numerous predecessors and studies also contributed. Groundwater modeling essentially began in the 1970s (Konikow and Bredehoeft 1974), although it too was based on predecessor work, primarily in the 1960s (Ogata and Banks 1961). Although it might be argued that many fundamentals in all this model pioneering were known well before the models themselves were established, these three examples should be considered the threshold events for formal modeling capabilities. Formal modeling might be considered as the representation of cause–effect relationships based on fundamental scientific principles in a way that is universal and reproducible, rather than anecdotal or empirical descriptions that might not apply universally.

Forty years ago, computerized modeling was as much a computer challenge as an environmental science challenge. For example, not only did the process of biochemical oxygen demand (BOD) need describing as an equation (e.g., Monod dynamics) with certain parameter values (e.g., the half-life), but the equations had to be programmed into a computer and synchronized with the other processes (e.g., reaeration). In those early days, we were still learning both the environmental science and the computer programming. Today, most of the environmental processes have been described, although they are still subject to improvement, and the model programs come on CDs that get loaded onto a PC (or via Internet downloads) and run via simple input screens with slick output graphs. Regardless of today's computer simplicity, however, good modeling still requires a good understanding both of the model and of the environmental processes.

In many cases, environmental models are used simply to examine resources. Examples include water supply, soil erosion, and river silting. Today's basic environmental models for *contamination* include:

- Stream models: Dissolved oxygen, an essential parameter for stream health, is often predicted in response to wastewater discharges, stormwater, photosynthesis, benthic demand, and reaeration (oxygen back into the stream from the atmosphere). Examples of such models are the QUAL series. These models can also be used to predict contaminant concentrations. Sediment transport in streams is often important because the sediments can carry contaminants; the US Army Corps of Engineers (USACE) Hydrologic Engineering Center (HEC) series of models is often used for this purpose.

- Air models: Downwind particulates, which can also carry contaminants, and vapor concentrations can be predicted with air models. As an example of air modeling challenges, a classic photo exists with plumes blowing in opposite directions from a tall stack beside a short stack. As noted above, the effects of both point sources and fugitive emissions are modeled, but the latter still present many problems in air modeling (see Chap. 10). Other important considerations for air models are terrain and the often highly variable direction and speed of the wind. Air stability is a gross but often useful input parameter. Common air models include versions of the Industrial Source Complex (ISC) models and the newly recommended (by the USEPA) AERMOD.

- Groundwater models: Both porous media (soils) and fractured flow (bedrock) models exist, and some attempts at karst (underground rivers) modeling are being made. A primary consideration is whether to model in two (lateral) or three (vertical also) dimensions, with the former being more common (and less expensive) after making simplifying assumptions. A key consideration for porous media models is how a contaminant might interact with the soils, often resulting in the use of retardation factors to account for slower contaminant flow than hydraulic flow due to the "chromatographic" effects of the soils. A common groundwater model today is MODFLOW.

- Soil vapor models: The entry into a home of volatile contaminants adsorbed to soils or dissolved in underlying groundwater is often modeled with the Johnson and Ettinger model (1991). Simply put, this model assumes that basements have a certain amount of cracks (a parameter input) and that volatiles in the subsurface soil pores are driven into the basement by pressure differences between the home and the subsurface after a volatile escapes from groundwater, according to Henry's Law, and desorbs from soils.

Many other models can be developed or customized for specific conditions. For example, air in bedrooms might be modeled after a prediction from the above-described model of vapors into the basement. Water column concentrations might be modeled after sediment transport is predicted. Breathing zone air might be modeled from emission predictions through the soil or other sources, such as a nearby river. Environmental professionals with a good understanding of the processes at work can model almost anything. The main consideration is whether to model or to measure.

When to Model, What to Look For?

Large data requirements for calibration and many hours to construct models with veracity are the primary considerations for deciding when to use a model. In other words, modeling can be expensive, so it better be worth it.

When is it worth it? When the results can be obtained no other way (e.g., predicting future conditions) and when the alternatives being predicted have large costs, such as the cost of being wrong or of an undesirable alternative. Avoid models that are more complex than they need to be. Be wary of certain models because they are always complex, such as fractured flow groundwater and sediment transport models.

Sometimes, modeling is simply a regulatory requirement. For example, modeling is often required for an air permit or to determine a stream's waste load allocation limits. In many cases, such modeling is more routine and less expensive.

Modeling has many inherent vulnerabilities and it is up to the modeler to prove these issues have been addressed properly, and show the possible shortcomings through a sensitivity analysis. For example, the vapor intrusion model noted above can give results varying by at least four orders of magnitude depending on what is specified for soil moisture content. Fractured flow groundwater models can be quite unreliable, depending on how the model is constructed. Unrealistic calibration values can make a model seem right but will result in very wrong predictions.

Flags to consider about a model's veracity include:

- Appropriate application: For example, sediment transport models might predict high flow (e.g., 100-year storm) impacts, but catastrophic events, which no model can yet predict, usually determine the important impacts (recall Hurricane Sandy).
- Appropriate construction: For example, are all the sources and processes accounted for?
- Data requirements: Since setting parameter values is critical to accurate modeling, consider if the available data are appropriate for establishing these values.
- Reasonable calibration: Consider if the parameter values in the calibrated model are within reason. For example, wind characteristics in an air model that differ significantly from the airport's wind rose should be suspect.
- Verification: Although not always possible due to data constraints, a verified model is much better than simply a calibrated model.
- Sensitivity analysis: Consider that both the important parameters have been identified and that their impact on the results has been examined by varying their values within possible alternative limits.

Environmental modeling can be an important tool, but it is expensive and must be done right, because it can be misleading.

References

Johnson PC, Ettinger RA (1991) Heuristic model for predicting the intrusion rate of contaminant vapors into buildings. Environ Sci Techol 25:1445–1452

Konikow LF, Bredehoeft JD (1974) Modeling flow and chemical quality changes in an irrigated stream-aquifer system. Water Resour Res 10(3):546–562

Ogata A, Banks RB (1961) Solution of the differential equation of longitudinal dispersion in porous media. U.S. Geol Surv Prof Paper 411-A:1–7

Streeter HW, Phelps EB (1925) A study of the pollution and natural purification of the Ohio River. Public Health Bulletin No. 146, U.S. Public Health Service

Turner DB (1961) Relationship between 24-hour mean air quality measurements and meteorological factors in Nashville. Tennessee. J Air Pollut Control Assoc 11(10):483–489

Turner DB (1967) Workbook of atmospheric dispersion estimates. U.S. Public Health Service, Public Health Service Publication No. 999-AP-26, Cincinnati, OH, 84 pp

Chapter 8
Risk Assessment

Abstract Human health risk assessment is useful to determine if an environmental condition is safe or permissible, and to determine appropriate cleanup levels. Risk assessment consists of two parts—exposure analysis and toxicity analysis (i.e., getting a chemical to the body and a chemical's health impact once it is in the body). Exposure analysis often requires further evaluation of existing data—for example, using soil contamination data to determine volatilization and breathing zone air concentrations. Toxicity analysis typically converts the exposed concentration to a dose and then compares that dose to reference material on safe doses, such as the US Environmental Protection Agency (USEPA) Integrated Risk Information System (IRIS) database. Risk assessment, as it is applied today on environmental problems like Superfund sites, could be improved by performing it as a risk–benefit analysis.

Keywords Hazard index · Cancer slope factor · Risk–benefit · Reasonable maximum exposure · Exposure point concentration · Superfund · IRIS

Introduction

A standardized form of risk assessment from chemicals in the environment has been performed routinely in Superfund decision making since the 1980s (USEPA 1989). Risk assessment is also commonly used in "toxic torts."

The National Research Council (NRC) defined risk assessment as having four parts (NRC 1983)—hazard identification, dose–response assessment, exposure assessment, and risk characterization—but it can be understood more simply as having two parts—exposure analysis and toxicity analysis—to the human body and in the human body. Many people view risk assessment as a black box and assume it is complicated, but it is much simpler than many people believe. This chapter attempts to unveil the mystery.

Superfund risk is expressed in one of two ways. It can be a probability of getting cancer from an exposure to carcinogens (i.e., 10^{-3} means a one-in-one-thousand incremental chance from a particular exposure over the one-in-three chance one already has in life of getting cancer) where the probabilities of multiple chemicals

N. Shifrin, *Environmental Perspectives,* SpringerBriefs in Environmental Science,   49
DOI 10.1007/978-3-319-06278-5_8, © The Author(s) 2014

are summed.[1] It can also be a ratio to the "safe" dose for noncarcinogens (i.e., a hazard quotient, HQ, of 2 means the exposure is two times higher than the safe dose, whereas the hazard index, HI, is the sum of HQs if multiple chemicals are present). Carcinogens can also contribute to noncarcinogenic health effects.

US Environmental Protection Agency (USEPA) has stated that a 10^{-6}–10^{-4} carcinogenic risk is the "target range" for managing Superfund risks, while risks after cleanup generally should offer no more than 10^{-6} cancer risks (USEPA 1991). It is sometimes unclear what this means; in practice, USEPA often requires cleanup actions whenever a 10^{-6} cancer risk or an HI of 1 are exceeded. For perspective, in-home radon risks are presumed acceptable if they pose 10^{-4} cancer risk (radon causes lung cancer), and an airplane crash offers about a 10^{-7} risk.

Two other key concepts for Superfund risk assessment are that "current or reasonably likely" property uses and a "reasonable maximum exposure" (RME) must be considered. These two concepts are commonly handled in risk assessments by assumptions of residential redevelopment and by the use of the 95 UCL for exposure concentrations, along with extreme exposure behaviors, such as very frequent visits.

These and other risk perspectives underscore one of the key flaws with Superfund risk assessment—it is one-sided and does not consider benefits. Many regulators and risk assessors do not embrace comparisons to daily risks, arguing that such daily risks are *chosen* by people because of their benefits, while people exposed to hazardous waste sites have no choice. Although this may be true for the decision for action (often based on a "baseline risk assessment"), each cleanup alternative has different benefits and should be considered in terms of a risk–benefit analysis, commonly used by engineers and economists in other types of analyses. The National Contingency Plan's (NCP) nine criteria for evaluating action alternatives (Federal Register 1990) are perhaps an attempt at risk–benefit analysis, but the approach fails to have the computational veracity of a real risk–benefit analysis.

Exposure Analysis

A separate analysis of how people are exposed to chemicals is often the first step in risk assessment. For example, if a chemical is known to exist in a river, how much gets into fish tissue, and how much fish (from that river) does a person eat? If a chemical is in groundwater under a house, how much might volatilize into the upper soil level ("vadose zone") and seep into the home's basement or living area?

Sometimes an anticipated risk assessment can guide data collection, but sometimes it is less expensive or technically superior to estimate exposure concentrations from other measurements. For example, a site investigation might generate soil gas data, but breathing zone concentrations might be more pertinent for the risk assess-

---

[1] Another way to think about this is that there is one additional cancer from the exposure in a population of 1,000.

ment and thus must be estimated from the former. The actual collection of breathing zone air might be confounded by other sources, so such data would misrepresent the effect of the soil vapor.

Exposure analysis has three parts: (1) identification of exposure scenarios, (2) selection of exposure parameters and their quantification, and (3) estimation of exposure point concentrations (EPCs). Exposure scenarios should encompass all the worst—but still reasonable—current and future property uses by people, such that all other uses would pose less risk. For example, if a residential scenario is likely, examination of a weekend visitor is not necessary if a child and adult permanent resident scenario is already included.

Once appropriate scenarios are selected, exposure parameters must be assigned and quantified. *Assigned* would mean selection of "visits per year" as a pertinent parameter for (e.g., trespassing), while *quantified* might be an estimate of ten visits per year for that parameter. For example, how long will the adult be a resident? Will she dig in contaminated dirt, and if so, how much dirt will get on her skin? If contaminated dirt is on her skin, how much chemical will be absorbed through the skin? If contaminated dirt is on her hands, will she put her hands in her mouth and ingest dirt?

Exposure parameters depend on the scenario. Typical exposure scenarios are devised to account for a resident (child and adult), construction or utility worker, landscaper, trespasser, neighbor, worker, and "recreator" (e.g., swimmer), among others. USEPA publishes its *Exposure Factors Handbook,* which often simplifies identification and quantification of exposure parameters (USEPA 2011). Exposure parameters and their numerical values can have a significant effect on risk calculations. For example, an assumption that a trespasser encounters a contaminated property 5 days a week versus a more likely once a month increases the risk estimate by a factor of 20.

An EPC is another estimate that combines judgment with mathematics that can significantly affect a risk estimate. In theory, the math is straightforward (e.g., a 95UCL of the data is a well-defined concept), but of what data: of the dirtiest well with data collected over time, all wells (i.e., over space), or all "on-site" wells? Sometimes there are no data, and the EPC must be modeled or otherwise estimated. For example, surface soil data might exist, but the exposure scenarios might include a neighbor breathing dust blown downwind from the site. That single EPC might involve hundreds of hours to develop, calibrate, run, and describe an air model. Moreover, the wind varies, so what does that variation do to the neighbor's EPC? In regulatory risk assessment for Superfund, there is often a tension between being realistic and accounting for the "worst case."

An exposure analysis is only three steps—scenarios, parameters, and EPCs—but to perform it properly takes skilled scientists, good judgment, and an understanding of regulations and guidance.

Toxicity Analysis

Just as the exposure analysis offers many opportunities to overestimate risk, the toxicity analysis has an inherent bias for overestimating risk despite there being little leeway to vary this analysis. USEPA maintains a toxicity factor database called Integrated Risk Information System (IRIS; USEPA 2013), and it is difficult to use in a regulatory risk assessment any other toxicity factor values other than those in IRIS. Although they can be challenged, IRIS values are most commonly used, and a toxicologist's role in a risk assessment is often no more than looking up the values (assuming other kinds of scientists perform the exposure analysis). Toxicity factors, including those in IRIS, are often derived from laboratory animal data and thus include safety factors of ten or more to translate the tests to humans. These safety factors are the source of a possible upward bias of the toxicity assessment, although such factors may sometimes truly apply to the translation from rats to people.

In its simplest form, the toxicity analysis is limited to predicting the effect of a chemical once it is inside the body. However, issues of bioavailability can also be important. For example, certain chemical species of lead are not very bioavailable (i.e., they are typically excreted before causing harm to health), while others are quite bioavailable (i.e., the toxic effects of lead can be released upon the body). The uptake pathway can also be important. For example, the inhalation of fine lead particles can have more impact than dirt-bound lead on the skin. Sometimes the precise form of a molecule that exists as a group of related compounds can be important. For example, current theory is that "dioxin-like, coplanar" polychlorinated biphenyl (PCB) congeners are the toxic ones.

The toxicity analysis can sometimes be more complex than looking up IRIS values. If there is reason to challenge an IRIS value (the database is not static; new toxicity information is always appearing in the scientific literature), for example, in a toxic tort rather than in a Superfund remedial investigation (RI), a skilled toxicologist is invaluable. It is also important to be aware that there are at least two types of toxicologists—clinical and what might be considered "regulatory." The former, who might treat individual sick patients, are typically not the type to perform risk assessments.

What It Means?

The first lesson about risk assessment is to understand how it can overestimate risk. The word "reasonable" is thrown around in risk assessment like a tennis ball to a dog, but is any particular risk assessment reasonable? As shown above, there are many opportunities to overestimate risk in the exposure analysis, and the toxicity analysis may often have an inherent upward bias. The net result is that Superfund risk estimates may overestimate risk by one or more orders of magnitude.

That is the second lesson of risk assessment—it is an "order-of-magnitude" science. Thus, seemingly precise results, such as a 3.45×10^{-6} incremental cancer risk, are quite meaningless, yet they are used as bright lines for decision making. An uncertainty analysis is a required part of Superfund risk assessments, but it is often rote and rarely used for anything. People at or near Superfund sites should not get too worried when risk estimates are slightly above action thresholds (which often remain unclear) and might only start paying attention when the results are orders of magnitude above. "Paying attention" should start with an examination of the scenarios considered in the risk assessment and comparison to one's actual exposure scenario. The numbers too often take on a life of their own, and common sense suffers.

Another potential flaw with risk assessments is the estimation of noncarcinogenic risks. Noncancer risk reference doses (RfDs) derive from specific health impacts. For example, an RfD for kidney failure from lead exposure might be available from a rat test. As noted above, HQs are the ratio of the actual dose to the RfD, and the HI is the sum of HQs. But it is only proper to sum HQs having the same health effect "endpoint"—for example, only the HQs that cause kidney failure. The HI is sometimes calculated incorrectly by summing all HQs, regardless of endpoint, which results in an overestimate of noncancer risk.

Perhaps the most important lesson about risk assessment pertains to the response decision making. As noted in the introduction, the "baseline risk assessment," which is an evaluation of current and reasonably foreseeable exposures, might stand on its own, but consideration of responses should be based on a risk–benefit analysis. For example, PCBs in the Hudson River were predicted to decline over the coming decade to "safe" levels, but the current risks exceeded action thresholds, so remediation was required. What is the benefit of spending now what will become billions of dollars for dredging versus waiting for the decline? The benefits of dredging now could be monetized in terms of the value of lost fishing, water supply treatment costs, and other impacts, but this was not done in a traditional risk–benefit analysis before the decision was made.

Despite its vulnerabilities, the most valuable aspect of risk assessment is that it provides a rational tool where the alternatives are likely to be more subjective. However, common sense must prevail in the analysis, and the tendency of risk assessment to overestimate risk must be considered in a more meaningful way than the currently typical uncertainty analysis. Using regulatory guidelines, one exercise to calculate when to leave for the airport concluded that 9 h would be best (HWCP 1993). Superfund response decision making should be based on conventional risk–benefit analysis, not simply on a risk analysis, while applying the NCP's nine criteria.

References

Hazardous Waste Cleanup Project (HWCP) (1993) Exaggerating risk. Hazardous Waste Cleanup Project, Washington, DC

National Research Council (NRC) (1983) Risk assessment in the Federal Government: managing the process. National Academy Press, Washington, DC

USEPA (1989) Risk assessment guidance for superfund, 4 vols. EPA 540/1-89/002

USEPA (1990) 40 CFR Part 300: national oil and hazardous substances pollution contingency plan; Final Rule. Fed Regist 55(46):8666–8865

USEPA (1991) Role of baseline risk assessment in superfund remedy selection decisions. OSWER Directive 9355.0-30

USEPA (2011) Exposure factors handbook: 2011 edition. EPA 600/R-090/052F

USEPA (2013) Integrated Risk Information System (IRIS). www.epa.gov/iris. Accessed 15 June 2013

Chapter 9
Water Quality and Its Management

Abstract Water quality is judged by water's chemical constituent concentrations and general quality parameters, like dissolved oxygen and suspended solids, against water quality criteria established for the intended use of the water, which can vary from water supply to industrial discharge conveyance. Water quality management is the control of discharges, both point and nonpoint sources, to the extent necessary to maintain intended water quality criteria. Wastewater treatment uses various approaches and control technologies, depending on the nature of the wastewater and the degree of required treatment, to attain the control required to ensure the intended receiving water quality. Water quality in the USA is managed primarily through Clean Water Act regulations, which include permit requirements for all discharges.

Keywords Water quality · Wastewater · Classification · Waste load allocation · TMDL · NPDES · Secondary treatment · Pollutants · SPCC · Leaks · Spills · BOD

Introduction

To some degree, judging the quality of water depends on its intended use. For example, treated sanitary sewage might be acceptable for irrigation but not for drinking. State classification systems usually define water's intended use. Water is a precious resource, as is well known by the three billion people who spend more than 4 h a day obtaining it. As natural resources become scarcer, the preservation and restoration of water quality will become more important. This will require an uncomfortable balance with another resource—energy. Water and wastewater treatment requires considerable energy. There will also be more recycling, although much subtle recycling already exists, such as by the thousands of water intakes located downstream of wastewater outfalls.

N. Shifrin, *Environmental Perspectives*, SpringerBriefs in Environmental Science,
DOI 10.1007/978-3-319-06278-5_9, © The Author(s) 2014

Definitions and Measurement of Water Quality

The ability to measure water quality parameters, as well as prevailing concerns, somewhat determines how water quality can be defined. For example, until the advent of instrumental analytical chemistry in the 1970s, limited wet chemistry methods generally confined the definition of water quality to the "conventional parameters": primarily biochemical oxygen demand (BOD), dissolved oxygen (DO), pH, suspended solids, bacteria, nutrients, and a few trace metals, like lead. In general, these parameters were sufficient to define water quality when most discharges were highly attenuated by "self-purification" (i.e., natural recovery) and before large populations and extensive industrialization demanded more attention. Other conventional parameters such as alkalinity, hardness, and total dissolved solids were also concerns for certain water supply needs. Fish kills and turbid water, often measured with a Secchi disk, were among the most pronounced water quality problems, historically.

Lower detection limits and the ability to measure many more compounds with laboratory instruments like atomic absorption spectrometer (AAS) and gas chromatograph/mass spectrometer (GC/MS) allowed more elaborate definitions of water quality starting in the 1970s. This timing aligned with US Environmental Protection Agency's (USEPA's) first definition of 64 priority pollutants in 1976 (it was actually 65, but dioxin was sidelined), which initiated a more chemical-specific definition of water quality. Although they do not apply to ambient waters, maximum contaminant levels (MCLs) developed under the Safe Drinking Water Act in the late 1970s served as one benchmark for high-quality water in terms of specific chemicals. USEPA also published Ambient Water Quality Criteria in 1968, 1980, and 1998. Before that, very few agency-based water quality standards or criteria existed. The first water quality standard, issued by the US Treasury in 1914 attempting to control diseases accompanying shipped imports, involved a single parameter—bacteria. As late as 1946, there were only seven water quality criteria: four trace metals (arsenic, lead, mercury, and chromium), one poor measure of organic compounds (phenolics), suspended solids, and bacteria (Shifrin 2005).

The 22 editions of *Standard Methods for the Examination of Water and Wastewater,* first published in 1905, chronicle the evolution of both the ability to measure water quality and its definition. More recently, USEPA has published several measurement manuals (cf. USEPA SW-864 and its Contract Laboratory program requirements). Hundreds of individual chemicals can now be measured in water, some at extremely low detection limits such as 1 pg/L.

Because pristine water everywhere is unlikely and unnecessary, ambient waters are assigned quality classifications. State agencies define these classifications based on desired quality and pragmatic needs for wastewater disposal. Nomenclatures vary by state but generally are alphabetic, with Class A waters suitable for drinking, Class B suitable for contact recreation, down to Class E in some cases, suitable for only wastewater conveyance. Classifications provide a goal for water quality management and a basis for discharge permits and waste load allocations, described below.

In addition to their technical role, water quality parameters also provide a conceptual understanding of water quality. The conventional parameters still play a central role for this understanding:

- *BOD* is a measure of organic matter decay, which exerts a drain on oxygen dissolved in water.
- *DO* is a measure of water's ability to support life. The solubility limit of oxygen in water is a function of temperature and is about 10 mg/L at room temperature. Trout are very sensitive to DO, and a minimum of 5 mg/L DO is generally targeted for trout streams, which are often viewed as pristine.
- *Suspended solids* are a measure of water clarity, sometimes also measured as turbidity (a measure of both suspended solids and particle size), and are both an aesthetic parameter and a functional one, because some water uses cannot tolerate high suspended solids.
- *Bacteria levels,* measured by culturing coliform bacteria from a drop of water, are used as a surrogate for the pathogenic disease potential of water. The theory is that coliform bacteria residing in the intestines of warm-blooded animals might represent the spread contagious diseases.
- *Nutrients* (e.g., nitrates and phosphates) are a water quality concern because excessive nutrients can allow excessive microorganism growth (e.g., algae blooms), which are not aesthetic and can deplete DO when they die and decay.
- Individual chemicals are generally a toxicity concern to humans, fish, or both. For example, some organic chemicals and arsenic cause cancer after prolonged exposure, while some trace metals cause certain diseases, like lead retarding brain development in children and causing hypertension in adults.

Sources of Water Pollution

The three primary sources of water pollution are municipal wastewaters, industrial wastewaters, and stormwater. Spills, leaks, and mining waters also can be important. The 1972 Clean Water Act (CWA) revolutionized how these sources are managed in the USA, as discussed in the last section of this chapter.

Municipal Wastewaters

Municipal wastewaters are usually considered in terms of their sanitary, organic, and particulate nature. Many texts have been published on this topic, and the Water Environment Federation publishes several *Manuals of Practice* on this and related topics, such as stormwater control. The quality of municipal wastewater might be understood by considering its sources: residential wastewaters, including sanitary and cleaning; excess water (e.g., drips and faucet drainage); commercial wastewaters, such as from restaurants; and some industrial wastewaters that discharge into municipal sewers. Groundwater inflow to sewers can also be a concern, sometimes contributing up to 10% of flow.

A rule of thumb for municipal wastewater flow is about 150 gallons per person per day, although conservation efforts will influence this rule. The primary quality view of municipal wastewater is BOD, bacteria, suspended solids, and nutrients, although the industrial component in some municipal wastewaters can add other concerns. Industrial discharges to municipal systems might cause two concerns: interference with treatment and treatability/discharge.

Industrial Wastewaters

Industrial wastewaters are discharged either directly to receiving waters or indirectly via municipal sewer discharges. The most important consideration about industrial wastewaters is their variability. The myriad of industry types and variations within an industry and even within a plant defies a simple definition. Many textbooks exist about industrial wastewater, and USEPA has published many studies by industry. Two common steps in industrial wastewater control are suspended solids removal and pH adjustment. Beyond that, it depends on the industry. For example, chemical plants often pose specific chemical considerations, metal plating poses trace metal issues, petroleum refineries have oily wastewaters, and so on. Close scrutiny of each plant process is often required to understand how and what wastewater is generated in order to understand its wastewater issues.

Stormwater

Stormwater can pose significant potential pollution and has the added issue of high-flow variability. When the CWA was first passed, many believed stormwater was a minor issue compared to wastewater, but this has often been disproven for two reasons: (1) it is more contaminated than originally believed and (2) it plays a more significant role as higher standards for receiving water quality evolved. Today, Total Maximum Daily Load (TMDL) studies are required to understand both wastewater and stormwater loads to receiving waters. Stormwater can be discharged directly to receiving waters, as runoff, through storm sewers that drain runoff, and through combined sewer systems (i.e., where trunk lines connect both sanitary and storm sewers). The combined sewer issue is particularly problematic because storm surges flush into receiving waters both sanitary and contaminated stormwater, as well as highly concentrated sediment that collects at the bottom of sewer pipes during lower-flow, dry periods. Stormwater typically is high in suspended solids, but it also contains whatever the runoff collects from the ground surface, including BOD and chemical contaminants.

Leaks, Spills, and Mining Waters

Leaks and spills occur wherever fluids are handled. In 1975, about 20 million gallons of fluids accidentally leaked or spilled in the USA, 69% from equipment failures (Lindsey 1975).

Ore mining and milling use tremendous amounts of water. Acid mine drainage, created by bacterial action on pyrite, which is common in mines, can cause significant water quality problems. Thermal pollution, such as cooling water from power plants, can also cause water quality problems when the heat alters DO levels and living conditions for aquatic life.

Essentially, all wastewater is treated and managed before being discharged, as discussed in the next section. Treated wastewaters are typically discharged to streams and lakes, but in coastal areas discharges to the ocean or estuaries are also common and about 10% of wastewater is injected into the ground.

Wastewater Treatment

Wastewater treatment methods are considered as primary, secondary, tertiary, and advanced. Each is additive upon the prior one. Primary treatment is essentially solids removal. Secondary aims at BOD removal. Tertiary usually aims at nutrient removal. Advanced aims at specialized issues, such as specific chemical contamination, or at very high degrees of municipal wastewater treatment for recycling purposes.

Many treatment systems work best under constant flow conditions, which is difficult to maintain in many cases, such as under diurnal variations in municipal wastewater flow (residents use little water while sleeping), and surge flows during storms. Many treatment systems have flow equalization tanks to deal with this issue. A typical treatment plant has many appurtenances to the main units noted above, such as grit screens, chemical mixers, flocculators, sludge treatment, and now often chlorinators. Many textbooks provide details on wastewater treatment design (Kolarik and Priestley 1996; Tchobanoglous et al. 2003; Surampalli and Tyagi 2004; Javid and Khan 2011; WEF 2009; Daigger and Love 2011; Henze et al. 1997; Gray 2004). In addition, the classic *Water and Wastewater Engineering* still offers tremendous insight today for both water quality and wastewater treatment issues (Fair et al. 1968).

Municipal wastewater treatment plants often have a treatment train that includes large solids removal (e.g., bar screens), primary sedimentation, biological treatment (e.g., activated sludge), secondary sedimentation, nutrient removal, and chlorination. Sludge generated from the sedimentation steps is often digested (anaerobic biodegradation), with the methane used in the plant for energy and the reduced solids dewatered and disposed, often by landfilling. Municipal effluent and sludge might have specific chemical contamination issues depending on what is discharged into publicly owned treatment works (POTW) sewers. Particularly in arid climates, municipal sewage might be treated further, such as with reverse osmosis, so that the effluent can be recycled. Some POTWs in those situations even separate sanitary sewage from "greywater" (e.g., laundry water) so that the latter can be treated differently and recycled.

Industrial wastewater treatment methods vary considerably by industry. Many use flow equalization, sedimentation, and neutralization (pH control, which can result in the need for more solids removal). After that, it depends on the industry and is plant-specific. In general, methods for trace metal removal include coagulation/flocculation/sedimentation using alum (aluminum sulfate), ferric chloride, and polyelectrolytes to enhance coagulation, or ion exchange columns packed with various materials that exchange "good" ions for "bad" ions with periodic regeneration with waste concentrates treated further or disposed. Membranes, which can be made to selectively allow certain ions to pass through, are sometimes used (Kislik 2009). Reactive treatment also may be used, such as for the conversion of toxic hexavalent chromium to less toxic trivalent chromium. Destruction is also sometimes used, such as the oxidation of cyanides, which are often used in electroplating.

Organic chemicals are often treated with activated carbon (Cecen and Aktas 2011; Bansal and Goyal 2007, 2010), most often with a granular form in packed columns (GAC, granular activated carbon), or by adding it as a powder to wastewater, with subsequent removal by sedimentation (PAC, powdered activated carbon). Advanced filtration methods are sometimes necessary, such as mixed media bed filtration, reverse osmosis (also removes certain dissolved contaminants), or ultrafiltration (AWWA 2007; Kucera 2010; Wankat 2011).

Stormwater is typically treated with some kind of sedimentation device (NRC 2009; Debo and Reese 2002; Erickson et al. 2013). The challenge is to allow efficient removal while accounting for flow surges and even prolonged dormant periods. Many stormwater treatment units are rudimentary, such as the surge basins appearing along many highways today. Basins or lagoons have been a longstanding wastewater treatment approach, with about 180,000 used in the 1980s (USEPA 1983).

One important consideration for wastewater treatment is residual generation. Very few treatment methods actually destroy contaminants. More often, they concentrate them and sometimes change their form. These residuals sometimes are treated further, but often require disposal. That is why the term "waste management" is more applicable—there is no free ride.

Water Quality Regulation

The 1972 federal CWA revolutionized water quality management in the USA. Significant requirements of the CWA include:

- *Wastewater discharge permits*. The CWA made it illegal to discharge from any pipe in the USA without a permit. USEPA or state designees administer this permit program via the National Pollutant Discharge Elimination System (NPDES) or the state equivalent (e.g., New York calls it SPDES for State Pollutant Discharge

Elimination System). The first NPDES permits, focusing on wastewaters, were issued in 1974, while a stormwater permit program began in the 1990s. Permits contain limits for pertinent individual parameters, a monthly average and maximum, and flow, usually in terms of both concentration and load (e.g., pounds), as well as monitoring requirements. Although limits in many permits have declined over the years as treatment technologies and water quality expectations have improved, most permits still do not require zero discharge. Permits are generally renewed on a 5-year cycle.

- *Wastewater treatment.* The CWA required that all wastewaters be treated to increasingly more stringent degrees in two initial steps: best practicable technology currently available (BPT) by 1977 and best available technology (BAT) by 1983. USEPA defined what these terms meant for each discharge. Although the deadline for these goals have long past and were generally met (some coastal cities resisted, arguing that the ocean provided sufficient dilution), wastewater treatment continues to improve as technology advances and expectations rise.
- *Spill control.* The CWA required that entities subject to spills and leaks, usually industry, develop Spill Prevention, Control, and Countermeasure (SPCC) plans. These plans include "hardware," like dikes around tanks and drip pans under pumps, and also management procedures, such as what to do when a spill occurs.
- *Toxic pollutants.* Section 307 of the CWA required USEPA to regulate specific chemicals. The agency was sued in 1976 to enforce this requirement. The result was the agency's issuance initially of a list of 65 priority pollutants, later expanded to 129.

Many other regulatory elements of water quality management exist in the USA. Some are administered by USEPA and others by the states, as long as states can demonstrate their requirements are at least as stringent as the federal ones.

Another important element of water quality management and regulation is waste load allocation. This had been practiced by sanitary engineers and agencies well before the CWA, but it became an important tool to implement NPDES, and now all discharges, by morphing into TMDL analysis, the modern version of which focuses on "impaired waters" (i.e., waters that do not meet their classifications). The basic concept of waste load allocation (now TMDL analysis) accounts for point and nonpoint discharges to a receiving water to determine the allowable load from each that will enable the receiving water to meet its water quality classification. This often uses water quality modeling as a tool to iterate the waste loads until desirable water quality is predicted (Benedini and Tsakiris 2013). Many software packages exist to perform such modeling conveniently, although the modeler must understand many water quality issues. Water quality modeling derives from oxygen balance models developed by Streeter and Phelps (1925) of US Public Health Service (USPHS).

The "allowable" load from each discharge is often a tenuous balance between what is feasible and what is necessary. Depending on the water's classification/goal, it is often necessary to "push the envelope" to set increasingly stringent goals for allowable loads. As we set increasingly more stringent requirements and goals, we must realize that we pay for this one way or another with money, energy, taxes, and the price of goods. We must weigh this cost against the benefits of a cleaner water environment.

References

American Water Works Association (AWWA) (2007) Reverse osmosis and nanofiltration. AWWA, Denver, 226 p
Bansal RC, Goyal M (2007) Activated carbon adsorption. CRC Press, Boca Raton, 497 p
Bansal RC, Goyal M (2010) Activated carbon adsorption. CRC Press, Boca Raton, 520 p
Benedini M, Tsakiris G (2013) Water quality modelling for rivers and streams. Water science and technology library, vol 70. Springer, Dordrecht, 288 p
Cecen F, Aktas Ó (2011) Activated carbon for water and wastewater treatment. Wiley-VCH, Weinheim, 406 p
Daigger GT, Love NG (2011) Biological wastewater treatment, 3rd edn. CRC Press, Boca Raton, 991 p
Debo TN, Reese AJ (2002) Municipal stormwater management, 2nd edn. CRC Press, Boca Raton, FL, 1176 p
Erickson AJ, Weiss PT, Gulliver JS (2013) Optimizing stormwater treatment practices: a handbook of assessment and maintenance. Springer, New York, 337 p
Fair GM, Geyer JC, Okun DA (1968) Water and wastewater engineering. Water purification and wastewater treatment and disposal, vol 2. Wiley, New York, NY
Gray NF (2004) Biology of wastewater treatment. Imperial College Press, London, 1421 p
Henze M, Harremoes P, Arvin E, LaCour Jansen J (1997) Wastewater treatment: biological and chemical processes, 2nd edn. Polyteknisk Forlag, Lyngby, 384 p
Javid S, Khan SJ (2011) Wastewater treatment and reuse: MBR approach: influence of nitrogen loading rate (nlr) on nutrients removal in Attached Growth—Membrane Bioreactor (AG-MBR). VDM Publishing, Saarbrücken, 96 p
Kislik VS (2009) Liquid membranes: principles and applications in chemical separations and wastewater treatment. Elsevier Science Ltd, Oxford, UK, 445 p
Kolarik LO, Priestley AJ (1996) Modern techniques in water and wastewater treatment. CSIRO Publishing, Collingwood, 200 p
Kucera J (2010) Reverse osmosis: design, processes, and applications for engineers. Wiley, Hoboken, 393 p
Lindsey AW (1975) Ultimate disposal of spilled hazardous materials. Chem Eng 23:107–114
National Research Council (NRC), Committee on reducing stormwater discharge contributions to water pollution (2009) Urban stormwater management in the United States. National Academies Press, Washington, 587 p
Shifrin NS (2005) Pollution management in the twentieth century. J Environ Eng 131(5):676–691
Streeter HW, Phelps EB (1925) A study of the pollution and natural purification of the Ohio River. U.S. Public Health Service Public Health Bulletin No. 146. February
Surampalli RY, Tyagi RD (2004) Advances in water and wastewater treatment. ASCE Publications, Reston, 585 p
Tchobanoglous G, Burton FL, Stensel HD (2003) Wastewater engineering: treatment and reuse. McGraw-Hill Education, New York, 1819 p
USEPA (1983) Surface impoundment assessment national report. Office of Water, Office of Drinking Water, Washington, DC. December. 224 p
Wankat PC (2011) Separation process engineering: includes mass transfer analysis, 3rd edn. Prentice Hall, Upper Saddle River, 955 p
Water Environment Federation (WEF) (2009) Design of municipal wastewater treatment plants MOP 8, 5th edn (WEF manual of practice 8: ASCE manuals and reports on engineering practice, No. 76). WEF and the American Society of Civil Engineers/Environmental and Water Resources Institute. Alexandria, VA, 2600 p

Chapter 10
Air Quality and Its Management

Abstract Air quality is difficult to measure accurately and is usually considered in terms of particulates and vapors. The United States Environmental Protection Agency (USEPA) developed six key "criteria pollutants" in the 1970s: particulates, sulfur oxides, nitrogen oxides, ozone, carbon monoxide, and lead. Some of these compounds react in the atmosphere to create smog. More recently, USEPA has established 189 hazardous air pollutants (HAPs). Air is polluted by stationary sources (point sources, i.e., stacks, and fugitive sources, e.g., blowing dust) and mobile sources (e.g., vehicles). Fugitive emissions have been the most difficult to characterize and control. Air quality within the USA is managed primarily through Clean Air Act regulations, which include a unique state–federal partnership for permitting, called State Implementation Plans (SIPs). Air emission controls are based primarily on filtration (e.g., baghouses), scrubbers, and management techniques, such as the use of cleaner fuels and fugitive source management techniques (e.g., wetting, covering, enclosing, and sweeping).

Keywords Air quality · Emissions · Particulates · HAPs · HiVol · Gaussian plume · SIP · Fugitive emissions · Point source · Mobile source · PM$^{2.5}$ · Baghouse · Scrubber · Electrostatic precipitator

Introduction

One of the most difficult issues for air quality and its management is its scale. Air emissions are massive, while the atmosphere offers a large amount of dilution and the sun offers strong degrading radiation. That same sun, however, can create new air quality problems, such as smog and ozone. Although atmospheric air is highly regulated in the USA, air pollution problems still exist. Indoor air can also be an issue, but it is regulated only in factories by the Occupational Safety and Health Administration.

N. Shifrin, *Environmental Perspectives,* SpringerBriefs in Environmental Science,
DOI 10.1007/978-3-319-06278-5_10, © The Author(s) 2014

Definitions and Measurement of Air Quality

For many years, air quality was considered only in terms of particulates (smoke and dust), although local conditions such as smelters caused early concerns about lead and sulfur dioxide. Air quality today is generally considered in two forms: *particulates* and *vapors*. Particulates can be simple dust or can be chemical-laden, such as lead oxide. Vapors can be toxic and reactive to cause new problems. The photochemical reactions of nitrogen and carbon compound emissions that create smog were first described in the 1950s (Haagan-Smit 1950). Smog (a contraction of smoke and fog) has sometimes been deadly, such as in Donora, PA, in 1948 and in London in 1952, and it is persistently troublesome in high-traffic areas subject to atmospheric inversions, such as in California.

To a large degree, air quality is defined by the way it is regulated. The United States Environmental Protection Agency (USEPA) has defined six criteria pollutants (particulates, sulfur oxides, nitrogen oxides, ozone, carbon monoxide, and lead) and 189 hazardous air pollutants (HAPs; e.g., benzene, chloroform, pesticides), which tend to be the current concerns about air quality. More on air regulation is presented at the end of this chapter.

Air sampling and measurement are more difficult than water sampling because the sampling technique can alter the sample or not be representative. This is due partly to the scale and variability of air zones of interest and partly to the nature of what is measured. For example, shifting winds, birds and insects, and particle size distribution in the air can affect particulate sampling. Particles of any type can cause health effects, and their chemical makeup also can be a concern.

USEPA established the PM2.5 (particulate matter of 2.5 μm or larger) standard for air in 1997 (previously the focus was on PM10) because it is believed that such small sizes can penetrate deep into the lungs and lead to disease. Particulates are often collected by passing air through a filter, a process that might affect the chemical nature of particles, such as causing HAPs adsorbed to particles to volatilize. The workhorse of particulate sampling has been the high-volume (HiVol) sampler, often run for 24 h to collect a sample. Other methods include impingers and tape filters aimed at discrete particle sizes and instantaneous measurements. The dichotomous sampler was developed in the 1970s following research indicating that atmospheric particles commonly occur in two distinct size modes, often referred to as "fine" and "coarse." It uses a "virtual" impaction technique that eliminated a "particle bounce" issue that could affect representative sampling (USEPA 1977a). Currently, state-of-the-art, near-continuous instruments are available for particle-size distribution measurement (USEPA 2004).

Vapors are sampled by collecting a measured volume of air stored in a vessel, such as a Summa canister, to be shipped to a laboratory for analysis. The most difficult air sampling of all is posed by chemicals that exist in both vapor form and adsorbed to particles.

Air is typically sampled as a network covering various compass directions and distances to account for source and atmospheric variability. Such sampling usually is accompanied by measurements of meteorological conditions with a "Met" station, which samples wind speed, direction, and sometimes temperature. Air sampling can be more expensive than water sampling.

Air particulates were originally analyzed for their constituents by "wet chemistry" (Skogerboe et al. 1977; USEPA 1977a; Ludwig et al. 1965), such as by the colorimetric dithizone method. Today, AAS and ICP emission spectrometry are more accurate methods of choice (USEPA 1998). GC/MS can be used for organic analysis.

Sources of Air Quality Problems

Stationary air quality sources are often referred to as point sources and fugitive emissions. Mobile sources, such as automobiles, are also important. Emissions from these sources are often difficult to characterize accurately. For example, it can be challenging to sample a 200-foot-high stack with emissions varying with production rates; or ground-level, diffuse fugitive sources, such as windblown dust from piles, trucks, or factory units; or moving sources along a highway. Sometimes the emission source itself is sampled; other times, sampling in the near-field downwind provides a better measure.

Years ago, emission inventories of many types of sources were conducted (Larson et al. 1953; NAPCA 1969a; Southerland 2005). This led to the publication of standardized emission factors for industrial air pollutant sources by the United States Public Health Service (USPHS) in 1965, which was expanded in 1968 (USPHS 1968). Emission factors were sometimes based on only a single or a few facilities and were sometimes derived by analogy (i.e., one type of measured emission applied to another, unmeasured one), but such factors were still useful for characterizing sources without field data. Such characterization is useful for air modeling, which is often required for air regulation.

Point source issues are defined by the type of process emitting through them. For example, power plant stacks emit the products of combustion, such as particulates, carbon dioxide, carbon monoxide, sulfur oxides, nitrogen oxides, and water. Mobile sources used to be a significant source of lead emissions before leaded gasoline was banned (Hamilton et al. 1998; Bloomfield and Isbell 1933; Lewis 1985; Lippmann 1990; Needleman 1998; Hernberg 2000). They still present significant problems for smog precursors and HAPs, such as benzene. Point and mobile sources can be significant, but the most difficult sources for air quality management are fugitive emissions.

With some understanding of point sources by the 1970s, fugitive emissions became recognized as a frequent key element of air pollution. Fugitive emissions were believed to pose barriers against attaining newly developed air quality standards (Lillis and Young 1975; McCutchen 1976; USEPA 1977a, 1982). Lillis and Young (1975) and USEPA (1977a) noted that fugitive emissions could have a greater effect on air quality in close proximity to a source of air pollution than stack emissions, in part because fugitive emissions tend to originate and remain near ground level. The challenge entails both measurement and control.

Fugitive emissions remain the most elusive element of air pollution character-ization (USEPA 1993). In their seminal article, Lillis and Young (1975) named two categories of fugitive emissions: (1) industrial, process-related fugitives and (2) fugitive dust. They defined industrial fugitive emissions as "both gaseous and par-ticulate emissions that result from industrial related operations and which escape to the atmosphere through windows, doors, vents, etc., but not through a primary exhaust system, such as a stack, flue, or control system"; and defined fugitive dust emissions as natural or anthropogenic dusts (particulates only) made airborne by wind, human activity, or both (Lillis and Young 1975). USEPA (1977a) defined industrial process fugitive particulate emissions to include both fugitive emissions (industrial) and fugitive dust emissions that originated from within industrial facil-ity boundaries.

Due to the dispersed nature of fugitive emissions, their measurement is difficult, and few reliable data exist (Lillis and Young 1975; USEPA 1977a, 1979a). This lack of reliable data contributed to the omission of fugitive emission sources in the first nationwide efforts to control air pollution (Lillis and Young 1975; USEPA 1977a).

USEPA (1976a, b, c) outlined three basic approaches for the measurement of fugitive emissions: (1) quasi-stack, (2) roof monitor, and (3) upwind–downwind. However, none of these techniques was widely accepted as accurate (McCutchen 1976; USEPA 1976a, b, c). USEPA's manual provided criteria for selecting the most applicable of the three methods for a given set of conditions, as well as detailed sampling procedures (USEPA 1976a, b, c). However, neither this nor a 1980 update (TRC 1980) provided sufficient quantitative information for assessing the accuracy or reliability of measuring fugitive emissions.

Emission modeling was an alternative approach to measurement, but fugitive emissions modeling was limited by the lack of reliable emission factors, so the problem is circular. USEPA (1977a) noted fugitive emission modeling flaws due to several complicating factors, such as variable emission rates and the lack of detailed particle sizing data needed to model deposition. Fugitive emissions remain today as the most difficult air quality issue.

Air Emissions Treatment

The treatment of air emission sources is different for point sources and fugitive emissions. The former relies more on equipment technology, while the latter relies more on management practices. The four primary particulate control devices devel-oped in the twentieth century were cyclones (centrifugal separators), electrostatic precipitators (ESPs), wet scrubbers, and baghouses (fabric filters) (NAPCA 1969b; USEPA 1977b; Cooper and Alley 1994; EC/R Inc. 1998). ESPs and baghouses have been perhaps the most prevalent technologies (Stern 1968; NAPCA 1969b; USEPA 1977b, 1980, 1995).

By the mid-1970s, it had become apparent that point source controls alone would not achieve air quality standards for particulates. USEPA thus turned its attention to fugitive emission control (Lillis and Young 1975; USEPA 1977a). USEPA (1977a) blamed the lack of reasonably available control technology (RACT) on the difficulties with fugitive emission control. Several older documents by USEPA (1977a, b, 1979a, b) still basically represent the state of the knowledge on available control technologies for fugitive particulate emissions.

Retrofitting existing plants with fugitive emissions control systems was considered difficult due to space and operational limitations (USEPA 1977b, 1982), and few major advancements in fugitive emissions control techniques have been made. As noted by one consulting firm (EC/R Inc. 1998), "The most widely used methods of controlling process fugitives are local ventilation (e.g., hoods) and building enclosure/evacuation; paving of unpaved roads; eliminating, reducing, or managing truck transportation; and street cleaning are the most effective techniques to reduce fugitive dust emissions from roads."

Air Quality Regulation

The most appropriate way to regulate air quality was debated for many years, with arguments ranging from it being a local problem to it being a vast problem crossing state and even international borders. The federal Clean Air Act (CAA) Amendments of 1970 established a nationwide program for air regulation and management that persists today. US air pollution management involves a complex regulatory structure through air pollution control agencies at the local, regional, state, multistate, and federal levels (often within the shadow of international treaties).

Early air management focused on visible smoke and airborne soot through municipal ordinances. For example, Chicago established an ordinance in 1881, followed by Cincinnati, Cleveland, Pittsburgh, St. Louis, and St. Paul circa 1900. By 1912, 23 of the 28 US cities with more than 200,000 people had enacted smoke abatement programs (Reitze 1991). Some programs were based on an entire air quality basin, such as in the Los Angeles County Air Pollution Control District (Martineau and Novello 2004).

Federal interest in air pollution increased after a deadly smog incident in 1948 in Donora, PA. At that time, USPHS called for further air pollution research, and by 1955, the first federal air regulation legislation was issued (Reitze 1991; U.S. Congress 1959). This 1955 federal air law offered federal funds (US\$ 5 million annually) over a 9-year period to conduct scientific research on the causes of air pollution and to assist state research and training efforts. More extensive federal regulation was absent at this time because "unlike water pollution, air pollution…is essentially a local problem" (Eisenhower Administration Bureau of the Budget, cited in Reitze 1991). Responsibility for administering funding was vested with USPHS within the Department of Health, Education, and Welfare (DHEW).

Starting in 1963, the federal government issued a series of CAA amendments that have become the foundation of our present air quality management framework:

- 1963 (with the 1967 Air Quality Act): enforcement through requests for DHEW hearings and state participation; auto emission regulations; grants for agencies to set ambient air quality standards (AAQS). Air quality control regions (AQCRs) were established with an unfolding cornerstone of federal–state joint effort.
- 1970: strengthened federal role with responsibility and enforcement placed within the newly formed USEPA. Six criteria pollutants (and air quality standards for them), national AAQs (NAAQs), USEPA-approved State Implementation Plans (SIPs), stationary source emission standards (new source performance standards, NSPs), and HAP standards (national emission standards for HAPs, NESHAPs) were established. Through the 1970s, the states assembled their air quality control programs and upgraded their SIPs under the 1970 CAA.
- 1977: extended NAAQs attainment to 1987 and had states upgrade their SIPs, particularly for non-attainment areas using RACT and by imposing lowest achievable emissions rates (LAER). More stringent requirements for new sources in non-attainment areas and prevention of significant deterioration (PSD) in attainment areas using best available control technology. Required 5-year reviews of NAAQs for the criteria pollutants.
- 1990: stronger focus on NAAQs attainment in AQCRs via revised SIPs, particularly for ozone and carbon monoxide, was established. The amendments also specified 189 HAPs (USEPA had up to that time named only eight), with technology-based emission standards via maximum achievable control technology, accidental release prevention, mobile source regulation, acid rain, and stratospheric ozone.

Overall, the 1990 CAA established an ambitious agenda for USEPA, states, and the regulated community. Thousands of USEPA rulemakings have been necessary to implement the 1990 CAA, accompanied by many guidance documents and even more interpretations (Martineau and Novello 2004). The states' responsibilities are even more extensive than USEPA's responsibilities, in that the states must apply the policies, regulations, and guidance to individual air emission sources within their boundaries. And, of course, the regulated community must find a way to comprehend this complex body of regulations and laws, determine its impact on the affected businesses and organizations, and implement the required controls in a viable and effective manner.

The federally approved, state-developed SIP process is the cornerstone of air quality management in the USA. At the industrial level, it is based on a chain of events that starts with a plant offering a plan. It is often based on air modeling of anticipated emissions, negotiation, and state adaptation into its statewide plan, including other emissions such as from mobile sources and nonindustrial sources. Finally, USEPA must approve the SIP.

USEPA (1992) provided comprehensive rules for this federal–state effort. It focused primarily on the SIP submissions required for non-attainment areas (i.e., areas failing to meet air quality standards). It also laid out USEPA's interpretation of the New Source Review (NSR) provisions of the CAA, requiring states to submit SIP revisions in 1992 and 1993 conforming their NSR programs for non-attainment areas to USEPA's interpretations. It identified 31 major deliverables pertaining only to the ozone and carbon monoxide portions of the act, due from the states within 4 years of enactment to provide that all necessary SIP revisions be made and approved by USEPA within 6 years of enactment. The 6-year deadline was also the milestone by which ozone non-attainment areas in moderate or worse conditions had to achieve a 15% reduction in VOC emissions. This time frame illustrates optimistic expectations for the length of time it takes under the existing federal–state regulatory apparatus to determine specific air quality control policies and see resultant improvements in air quality.

References

Bloomfield JJ, Isbell HS (1933) The presence of lead dust and fumes in the air of streets, automobile repair shops, and industrial establishments of large cities. J Ind Hyg 15:144–149

Cooper CD, Alley FC (eds) (1994) Air pollution control. Waveland, Prospect Heights

EC/R Inc. (1998) Stationary source control techniques document for fine particulate matter. Submitted to USEPA, Air Quality Strategies and Standards Division, October

Haagan-Smit AJ (1950) The air pollution problem in Los Angeles. Eng Sci 14:7–13

Hamilton A, Hardy HL, Harbison RD (1998) Hamilton and Hardy's industrial toxicology. Mosby, Maryland Heights, MO. 682 p

Hernberg S (2000) Lead poisoning in a historical perspective. Am J Ind Med 38:244–254

Larson GP, Fischer GI, Hamming WJ (1953) Evaluating sources of air pollution. Ind Eng Chem 45(5):1070–1074

Lewis J (1985) Lead poisoning: a historical perspective. EPA J. http://www.epa.gov/history/topics/perspect/lead.htm. Accessed 26 Aug 2011

Lillis EJ, Young D (1975) EPA Looks at 'fugitive emissions'. J Air Pollut Control Assoc 25(10):1015–1018

Lippmann M (1990) Lead and human health: background and recent findings (1989 Alice Hamilton Lecture). Environ Res 51(1):1–24

Ludwig JH, Diggs DR, Hesselberg HE, Maga JA (1965) Survey of lead in the atmosphere of three urban communities: a summary. Am Ind Hyg Assoc J 26(3):270–284

Martineau RJ, Novello DP (2004) The Clean Air Act Handbook, 2nd ed. American Bar Association, Chicago, IL

McCutchen GD (1976) Regulatory aspect of fugitive emissions. In: Symposium on fugitive emissions measurement and control, May 1976, Hartford, CT. Report to USEPA, Office of Research and Development, National Technical Information Service (NTIS), EPA-600/2-76-246, NTIS PB-261955, September

National Air Pollution Control Administration (NAPCA) (1969a) Air quality criteria for particulate matter. Report to U.S. Department of Health, Education and Welfare, National Technical Information Service (NTIS), AP-49, NAPCA-Pub-AP-49, NTIS PB-190251, 211 p, January

National Air Pollution Control Administration (NAPCA) (1969b) Control techniques for particulate air pollutants. National Technical Information Service (NTIS), NTIS PB-190253, 241 p, January

Needleman HL (1998) Clair Patterson and Robert Kehoe: two views of lead toxicity. Environ Res 78:79–85

Reitze AW (1991) A century of air pollution control law: what's worked; what's failed; what might work. The Clean Air Act Emendments of 1990: A symposium overview and critique. The Environmental Law, 21 Envtl. L. 1549. 62 p

Skogerboe RK, Hartley AM, Vogel RS, Koirtyohann SR (1977) Monitoring for lead in the environment. In: Boggess WR, Wixson BG (eds) Lead in the environment: a report and analysis of research at Colorado State University, University of Illinois at Urbana-Champaign, and University of Missouri at Rolla. National Science Foundation. Foundation, Washington DC, pp 33–70

Southerland JH (2005) An abridged history of emission inventory and emission factor activities. In: USEPA (2005) 14th annual emission inventory conference: transforming emission inventories—meeting future challenges today, Office of Air Quality Planning and Standards, Las Vegas, 12–14 April 2005, 18 p

Stern AC (ed) (1968) Air pollution, Vol III, 2nd edn. Academic, New York

TRC Environmental Consultants, Inc. (TRC) (1980) Protocol for the measurement of inhalable particulate fugitive emissions from stationary industrial sources. Draft Submitted to USEPA, 113 p, March

US Congress (1959) Public Law 159, Chapter 360. An Act: To provide research and technical assistance relating to air pollution control. July 14. 322–323

USEPA (1976a) Technical manual for the measurement of fugitive emissions: upwind/downwind sampling method for industrial emissions. EPA-600/2-76-089a

USEPA (1976b) Technical manual for the measurement of fugitive emissions: roof monitor sampling method for industrial emissions. EPA-600/2-76-089b

USEPA (1976c) Technical manual for the measurement of fugitive emissions: quasi-stack sampling method for industrial emissions. EPA-600/2-76-089c

USEPA (1977a) Air quality criteria for lead. National Technical Information Service (NTIS), EPA/600/11, NTIS PB-280411. December

USEPA (1977b) Control techniques for lead air emissions. Vol I: Chapters 1–3. National Technical Information Service, EPA-68-02-1375, EPA-450/2-77-012-A, NTIS PB80-197544, 214 p, December

USEPA (1979a) Assessment of the use of fugitive emission control devices. EPA/600/7-79-045, NTIS PB-292748, 86 p, February

USEPA (1979b) Control of particulate emissions in the primary nonferrous metals industries—symposium proceedings. Office of Research and Development, Industrial Environmental Research Laboratory, National Technical Information Service (NTIS), EPA-600/2-79-211, NTIS PB80-151822, 515 p, December

USEPA (1980) Industrial process profiles for environmental use: Chapter 27. Primary lead industry. EPA-600/2-80-168, 75 p, July

USEPA (1982) Control techniques for particulate emissions from stationary sources—Vol 1 and 2. Office of Air Quality Planning and Standards, EPA-450/3-81-005a, EPA-450/3-81-005b, September

US Environmental Protection Agency (USEPA) (1992). 40 CFR Part 52: State Implementation Plans; General Preamble for the Implementation of Title I of the Clean Air Act Amendments of 1990. Fed Reg 57(74):13498–13570. April 16

USEPA (1993) A review of methods for measuring fugitive PM-10 emission rates. Office of Air Quality Planning and Standards, EPA/545-R-93-037, NTIS PB94-204203

USEPA (1995) EPA Office of Compliance Sector Notebook Project: profile of the nonferrous metals industry. Office of Enforcement and Compliance Assurance, EPA/310-R-95-010, 140 p, September

USEPA (1998) Locating and estimating air emissions from sources of lead and lead compounds. EPA-454/R-98-006, 384 p, May

USEPA (2004) Air quality for particulate matter. Vol I of II. National Center for Research Assessment, RTP Office, Office of Research and Development, EPA/600/P-99/002aF. October

USPHS (1968) Compilation of air pollutant emission factors. Bureau of Disease Prevention and Environmental Control, National Center for Air Pollution Control, 67 p

Chapter 11
The National Contingency Plan

Abstract The National Contingency Plan (NCP) is USEPA's road map for Superfund. Its intent is to ensure high-quality, consistent studies and remedies throughout
the USA. In addition to defining rules for studies, the NCP allows for two types of
cleanups: removal actions (faster, shorter, simpler parts of the total remedy) and
remedial actions (the full process, expected to lead to closure). Superfund allows for
costs to be recovered by the original performing party depending on cost-sharing
issues, but the law requires NCP consistency. The NCP delineates nine criteria for
selecting appropriate remedies.

Keywords NCP · Superfund · CERCLA · Criteria · Remedial action · Removal
action · RI/FS · Remedial investigation · Feasibility study · Operable unit · Record
of decision

Introduction

The National Contingency Plan (NCP) is Superfund's road map. It originated as
a regulatory framework for oil spill cleanup and has been revised many times, including a major revision and clarification in 1990 to adapt it to Superfund (USEPA
1990a). The NCP provides rules for the gamut of Superfund responses, from site
identification to final cleanup. As noted in the United States Environmental Protection Agency (USEPA)'s preamble to the 1990 NCP update, the NCP exists to
ensure "Comprehensive Environmental Response, Compensation, and Liability Act
(CERCLA)-quality cleanups" throughout the nation. The term, "CERCLA-quality
cleanups" means many things, including decision-making standardization, quality
of work, required steps, and opportunity for response cost-recovery challenges. This
chapter describes NCP fundamentals, without the regulatory jargon, and identifies
critical issues related to "CERCLA-quality" for response implementation and for
comparing actions to NCP requirements for cost recovery.

N. Shifrin, *Environmental Perspectives,* SpringerBriefs in Environmental Science,   71
DOI 10.1007/978-3-319-06278-5_11, © The Author(s) 2014

NCP Outline

The 1990 NCP has 68 sections under 10 subparts (USEPA 1990a). Some of the technically important ones are listed below:

Section 300.150: Worker health and safety
Sections 300.170/.180/.185: Federal, state, local, and nongovernmental participation
Section 300.405: Discovery
Section 300.410: Removal site evaluation
Section 300.415: Removal actions
Section 300.420: Remedial site evaluation
Section 300.425: Establishing remedial priorities (e.g., national priorities list, NPL)
Section 300.430: Remedial investigation/feasibility study (RI/FS) and remedy selection under remedial actions
Section 300.435: Remedial design/remedial action (RD/RA) and operation and maintenance (O&M)
Subpart F: State involvement
Subpart G: Natural resource damages
Subpart I: Administrative record

Additionally, the NCP delineates potentially applicable requirements for public participation, the implementation of which is usually the responsibility of the oversight agency:

Section 300.155: Public information and community relations
Section 300.415(n): Community relations during removal actions
Section 300.430(c): Community relations during RI/FS
Section 300.430(f) (2), (3), (6): Community relations during selection of remedy
Section 300.435(c): Community relations during RD/RA and O&M.

Oil spill responses are still guided by the NCP via subpart D, "operational response phases for oil removal."

There are two key technical concepts in the NCP: *studies* and *remedies*. USEPA offers a plethora of guidance for studies (USEPA 1988a, 1989a, 1991a, 2000a, 2002), memos about how to make remedy decisions (USEPA 1989b, 1990b, 1991a, b, 1993a, 1995, 1997a, 1999, 2000b, 2002), and memos about recovering costs (Porter 1988).

The NCP specifies two remedy approaches—remedial and removal actions. The latter can have time/cost constraints and entail simpler studies (USEPA 1993b; Luftig 2000) for simpler sites or simple parts of a site. After much criticism about CERCLA responses taking so long, USEPA developed its Superfund Accelerated Cleanup Model (SACM; Clay 2002), which intended to speed things up but maintain NCP consistency. Overlain on the NCP's concepts of removal versus remedial actions, SACM invented the concepts of early versus long-term actions. Regardless, the following is how the NCP addresses a Superfund site:

- Identification and listing
- Preliminary assessment (sometimes "paper studies," sometimes with data, sometimes integrated with portions of an RI under SACM)
- Removal versus RA and operable units (OUs) decisions (the NCP provides guidance on these procedural forks in the road)
  - RI/FS, if a remedial action
  - Engineering evaluation/cost analysis (EE/CA), if a removal action
- Remedy decisions
  - Record of decision (ROD), if a remedial action
  - Action memorandum (USEPA 1990c), if a removal action
- Remedy designs, sometimes with further data collection, often called an RD Study
- RA or removal action (i.e., the actual cleanup)
- Post-remedy activities, such as O&M, 5-year reviews, closeout reports, and delisting

Delisting from the NPL (Luftig 1996) is the holy grail of Superfund, but often years of O&M are required for groundwater "pump and treat" systems, and USEPA performs 5-year reviews for no-action decisions. USEPA considers a site that has only O&M requirements left to be "construction complete."

In its NCP promulgation notice of 1990, USEPA defined its perspective on "CERCLA-quality" cleanups of hazardous waste. According to USEPA, CERCLA-quality cleanups must:

- Be protective of human health and the environment
- Utilize permanent solutions and alternative treatment technologies or resource recovery technologies to the maximum extent practicable
- Be cost-effective
- Attain applicable or relevant and appropriate requirements (ARARs; USEPA 1988b)
- Provide for meaningful public participation

USEPA notes, however, that the NCP is not a cookbook. Variations of all elements are possible, depending on site/case conditions, but "substantial compliance" with the NCP is required.

Noncompliance with the NCP is the hook often pursued to dispute cost recovery, when either USEPA "sends the bill" to responsible parties or responsible parties attempt to recover costs from other responsible parties via Sections 107 and 113 of CERCLA. Despite the latitude offered by USEPA's non-cookbook warnings, NCP and Superfund sites are often so complex, technically, that they offer hope for finding some bone to pick about costs. However, it is often an uphill battle to prove noncompliance, because an agency most often has ordered the work, and courts typically decide that agency orders equal NCP compliance.

Two barbs on the hook often honed for demonstrating noncompliance with the NCP are "cost-effectiveness" and "arbitrary and capricious" actions. The NCP requires USEPA (or the state lead agency) to consider cost in its decision making, and USEPA guidance emphasizes cost-effectiveness (USEPA 1996). A response might not be cost-effective in many ways, such as redundant studies, excessive remedies, or failed remedies due to poor data or poor data interpretation; but if an agency required the work, courts typically decide it was consistent with the NCP.

Work done by USEPA must not be inconsistent with the NCP. Work done by potentially responsible parties must be consistent with the NCP. This opposite wording may represent the same requirement, except for the burden of proof, but it might also imply further deference to the agency for court consideration.

"Arbitrary and capricious" means doing something according to one's will or caprice and therefore implies an abuse of power. One USEPA attorney has argued that it is the *selection* of the remedy, not its *implementation,* to consider for an arbitrary and capricious test (Tucker 2007). Along those lines, one might consider, among other things:

- If the decision was based on a reasonable risk assessment or on ARARs—or was it arbitrary?
- If the decision considered appropriate options—or was it capricious?
- How the decision used the data—or whether it was consistent with the data?
- If a less costly response would achieve the same objectives—and, if so, what justifies the more costly approach?

One prevailing explanation for taking Superfund actions is that harm is "imminent and substantial," but it is almost always unclear if this is the case. For example, it is possible that very few landfills make people sick, whereas air pollution can offer obvious imminent and substantial danger. A better recognition of imminence and potency might help avoid the appearance and perhaps the reality of arbitrary and capricious Superfund response decisions. The "imminent and substantial endangerment" notion has been Superfund's wedge since the Love Canal frenzy in the 1980s.

Response Actions

As noted above, the NCP allows for two types of cleanup responses—removal or remedial actions. The latter has more "hoops" for data collection, data analysis (e.g., risk assessment), and remediation. Completing an RI/FS is a big deal, often costs millions of dollars, and takes years.

Many Superfund sites are divided into OUs, also an NCP notion, based on differing environments. An example would be the separation of sediment from landside issues into separate OUs. This usually requires two RI/FSs, thus adding costs and time, but is often a logical approach because OUs are most often so different.

The OU approach is also used to phase work. Once the agency selects a remedy concept (ROD or action memorandum), an RD (remedial design) follows that often requires more data in the form of an RD study. For example, RI/FS data might define areas to be excavated, but an RD study is required to collect more data to define "cut lines." After all the data are collected, there are often still surprises "in the field." Between the seemingly never-ending need for more data, surprises in the field, and other "scope creep" issues, Superfund sites often cost much more than predicted by their RODs (Oak Ridge Laboratory 1995; GAO 2010). Also, every step requires agency review and approval, so it is no wonder that the average time for a Superfund response completion can be about 20 years (GAO 1998).

To promote timely responses, NCP requirements are less involved for removal actions than for remedial actions, although the former must still be technically sound. According to the NCP, "[R]emoval actions shall, as appropriate, begin as soon as possible to abate, prevent, minimize, stabilize, mitigate, or eliminate the threat…" (Section 300.415(b)(3)). Some examples of appropriate removal actions given in the NCP (Section 300.415) include:

- Excavation of highly contaminated soils where such actions will reduce the spread of contamination
- Removal of drums or other containers where it will reduce the likelihood of spillage or exposure
- Destruction and removal of equipment or vessels "by whatever means are available" where it will reduce the likelihood of spillage or exposure
- Provision of site security measures

Originally, there were time and cost limitations for removal actions (12 months and US$ 2 million), but USEPA invented non-time-critical removal actions (USEPA 1993b; Luftig and Breen 2000) to avoid such limitations. A non-time critical removal action is based on an EE/CA, which can be thought of as "RI/FS light," while USEPA's decision document is called the action memorandum, which can be thought of as "ROD light."

Although these are the two primary response approaches foreseen by the NCP (studies are also responses), things can get tangled quickly. For example, the SACM overlay is intended to speed things up, but USEPA guidance for SACM still insists on NCP compliance, which tugs in the other direction. Also, when PCBs are involved, they are regulated under the TSCA but may still need Superfund remediation under CERCLA. Similarly, there can be a blurred line between CERCLA actions and RCRA actions, so USEPA offered specific guidance to integrate between the two authorities and avoid redundant work (Herman and Laws 1996).

Ultimately, the NCP requires "CERCLA-quality" cleanups that are consistently thorough, accurate, cost-effective, protective, and timely.

The Nine Criteria

The five statutory requirements of CERCLA are (simplified): (1) protect human health and the environment, (2) meet ARARs, (3) be cost-effective, (4) use permanent solutions, and (5) prefer treatment. The NCP translates this into nine criteria for considering remedy alternatives (USEPA 1990d):

- Threshold criteria
  - Protection of human health and the environment
  - Compliance with ARARs
- Balancing criteria
  - Long-term effectiveness and permanence
  - Reduction of toxicity, mobility, or volume through treatment
  - Short-term effectiveness
  - Implementability
  - Cost
- Modifying criteria
  - State acceptance
  - Community acceptance

Although these are good considerations, it often remains unclear how one alternative has been selected over another. A quantitative risk–benefit analysis for remedy selection might be a superior approach. USEPA provides guidance on how to consider these criteria (USEPA 1990d), which is structured around the headings noted above for them:

- Threshold criteria absolutely must be met.
- Balancing criteria are how the trade-offs are considered.
- Decisions can be modified if the state or public demand it convincingly.

Taken together, the nine criteria represent the NCP adoption of the CERCLA statutory requirements, but Superfund remedy selection can still be a mysterious process (USEPA 1997b).

References

Clay DR (USEPA) (2002) Guidance on implementation of the Superfund accelerated cleanup model (SACM) Under CERCLA and the NCP. OSWER Directive 9203.1-03

Fields T (US Environmental Protection Agency (USEPA)) (1999) Memorandum to Superfund National Policy Managers, Regions I-IX & Regional Counsels, Regions I-X re: Interim Policy on the Use of Permanent Relocations as Part of Superfund Remedial Actions. Office of Solid Waste and Emergency Response, OSWER Directive: 9355.0-71P, EPA 540F-98-033, PB98-963305. June 30. 10 p

Herman S, Laws E (USEPA) (1996) Memorandum to RCRA/CERCLA policy managers and the regions re: coordination between RCRA corrective action and closure and CERCLA site activities

Laws EP (USEPA) (1995) Memorandum re: land use in the CERCLA remedy selection process. OSWER Directive No. 9355.7-04. 25 May

Luftig S (USEPA) (1996) Memorandum re: removal of NFRAP sites from CERCLIS

Luftig S (USEPA) (2000) Use of non-time critical removal actions in Superfund

Luftig S, Breen B (USEPA) (2000) Memorandum re: use of non-time critical removal authority in Superfund response actions

Oak Ridge Laboratory (1995) Resource requirements for NPL sites. University of Tennessee, Joint Institute for Energy and Environment, Knoxville

Porter W (USEPA) (1988) Transmittal of the Superfund cost recovery strategy. OSWER Directive 9832.13

Tucker WC (2007) Inconsistency with the NCP under CERCLA: What does it mean? 9 VT. J. Envtl. L. 1. 19 p

USEPA (1988a) Guidance for conducting remedial investigations and feasibility studies under CERCLA. EPA/540/G-89-004. OSWER Directive 9355.3-01. October

USEPA (1988b) CERCLA compliance with other laws manual: interim final. EPA/540/G-89/006

USEPA (1989a) Risk assessment guidance for Superfund Volume I, human health evaluation manual (Part A), interim final. EPA/540/1-89/002. December

USEPA (1989b) CERCLA compliance with other laws manual. OSWER Directive 9234.1-02. August

USEPA (1990a) 40 CFR part 300: national oil and hazardous substances pollution contingency plan, final rule. Fed Reg 55(46):8666–8865

USEPA (1990b) Superfund removal procedures: action memorandum guidance. OSWER Directive 9360.3-01. September

USEPA (1990c) Superfund removal procedures: action memorandum guidance. OSWER Directive 9360.3–01

USEPA (1990d) A guide to selecting Superfund remedial actions. OSWER Directive 9355.0-27FS

USEPA (1991a) A guide to principle threat and low level threat wastes. Superfund Publication: 9380.3-06FS. November

USEPA (1991b) Guidance for performing preliminary assessments under CERCLA. EPA/540/G-91/013. Publication 9345.0-01A. September

USEPA (1993a) Guidance for evaluating technical impracticability of ground-water restoration. OSWER Directive 9234.2-25. September

USEPA (1993b) Guidance on using non-time critical removal actions under CERCLA. EPA540-R-93–057

USEPA (1996) The role of cost in the superfund remedy selection process. EPA 540/F-96/018

USEPA (1997a) Use of monitored natural attenuation at superfund, RCRA corrective action, and underground storage tank sites; OSWER Directive 9200.4-17, interim final. 8 Dec

USEPA (1997b) Rules of thumb for superfund remedy selection. EPA 540-R-97–013

USEPA (2000a) Close out procedures for national priorities list sites. OSWER Directive 9320.2-09A-P. EPA 540-R-98-016. January

USEPA (2000b) Data quality objectives process for hazardous waste site investigations. EPA/600/R-00/007. January

USEPA (2002) Guidance for quality assurance project plans. EPA/240/R-02-009, December

U.S. Government Accountability Office (U.S. GAO) (1998) Superfund: times to complete site listing and cleanup. GAO-T-RCED-98–74

U.S. GAO (2010) Superfund, EPA's estimated costs to remediate existing sites exceed funding levels. Report to Congress. GAO-10-380

Chapter 12
Technical Impracticability

Abstract Some environmental contamination simply cannot be cleaned up. US Environmental Protection Agency (USEPA) has rules for determining when this is the case (i.e., when technical impracticability precludes cleanup). In such cases, an alternative response must still be health protective and containing. One condition that often leads to a conclusion of technical impracticability is when dense non-aqueous phase liquids (DNAPLs, e.g., tarry, oily, or otherwise immiscible liquid chemicals) are present in the subsurface. When DNAPL is known to exist, it might be more efficient to declare technical impracticability and move to "plan B" rather than wasting time and money on a "plan A" cleanup destined for failure.

Keywords NAPL · Nonaqueous phase liquids · DNAPL · Containment · Remediation · TI waiver

Introduction

After 33 years of Superfund responses, it is clear that some hazardous waste problems cannot be "cleaned up," although they *can* be managed (USEPA 1998, 1999). Cleanup may be limited because a site is large (e.g., many mining sites) or because of chemical types (e.g., for some trace metal contamination). Often, however, cleanup is limited because of the presence of dense nonaqueous phase liquids (DNAPLs), described below. Although US Environmental Protection Agency (USEPA) has expressed a preference for treatment in its Superfund regulations and guidance (USEPA 1990, 1997, 2008), it also provides guidelines that allow a shift from treatment to management due to the technical impracticability (TI) of the former (USEPA 1993a). Most often, Superfund TI considerations pertain to the inability to achieve drinking water standards in groundwater. Many Superfund remedies under way today will eventually be reconsidered, because years of effort and monitoring data will prove that drinking water standards cannot be met. This chapter examines whether such wheel spinning might be averted by more realistic remedy planning at the onset.

Bryan S. Pitts of Berkeley Research Group, Waltham, MA assisted with preparation of this chapter.

N. Shifrin, *Environmental Perspectives,* SpringerBriefs in Environmental Science,   79
DOI 10.1007/978-3-319-06278-5_12, © The Author(s) 2014

Typical Problems, Appropriate Perspective

It is important to understand the difference between cleanup and management. Hazardous waste site *cleanup* usually results in contaminant levels low enough to satisfy environmental standards and unrestricted land uses. This is often described as meeting applicable or relevant and appropriate requirements (ARARs), a National Contingency Plan (NCP) requirement (USEPA 1993a). For example, groundwater might be removed from the ground, treated, and reinjected (pump and treat) until the aquifer meets drinking water standards, a common ARAR.

In contrast, a hazardous waste site *management* remedy often leaves contamination untreated or less treated, but sequesters it with coincident land-use restrictions to minimize risks. For example, an abandoned landfill might be remediated by leaving the waste in the landfill while containing it with a perimeter groundwater management system and protecting future exposures by enacting deed restrictions to limit future redevelopment of the landfill footprint.

USEPA remedy guidance emphasizes a preference for treatment (USEPA 1990, 1997) rather than management. However, many remedies are a combination of both. For example, a source area of highly contaminated soil might be excavated and treated, while a containment system may operate around a larger, less-contaminated area to manage residual soil contamination.

Some contamination problems are so large or complex that management is the only option. Examples include:

• Complex geology/hydrogeology (e.g., fractured bedrock, karst terrain, multiple aquifers, and low-permeability soils)
• Presence of nonaqueous phase liquid (NAPL, any fluid that is not water), especially DNAPL
• Remedy complexities or constraints (e.g., strong contaminant adsorption to soils, buildings in the way)
• Too many sources, particularly if they are not associated with the site
• Excessive contaminant areas, volumes, or depths

DNAPL is perhaps the most pernicious problem at Superfund sites. DNAPLs, such as coal tar or dry cleaner solvent, sink deep into the subsurface and can serve as a source of groundwater contamination by dissolution for thousands of years. Worse, DNAPL can be difficult to find in the subsurface and even more difficult to remove completely (USEPA 1993b). Even small, residual quantities of DNAPL remaining after removing the bulk of it can still cause significant groundwater contamination for thousands of years, thus defying remedy closure.

In contrast, light NAPL (LNAPL), such as gasoline or oil, floats on the water table, making it easier to find and collect. Some LNAPL might remain after the bulk is collected from the water table, such as in "smear zone" soils where the water table fluctuates due to seasonal or other influences, but smear zone residual can often be cleaned up because it is shallow or may attenuate over reasonable timeframes.

As of 2003, there had never been a Superfund site with DNAPL below the water table where drinking water standards had been achieved (USEPA 2003). Through April 2012, about 60 % of the 1,685 Superfund sites had DNAPL, but USEPA had granted only 91 TI waivers at 85 sites (5 %; USEPA 1993b). Only 43 of those waivers (3 % of Superfund sites) were because of DNAPL (USEPA 2012, 2013a). Either the other 950 + DNAPL sites are actually not as hopeless for cleanup as DNAPL characteristics would imply or much money is being spent unnecessarily. If the latter, could time and money be saved by issuing TI waivers at the beginning?

In its TI waiver guidance, USEPA (1993a) says that it "may be difficult to determine whether cleanup levels are achievable at the time a remedy selection decision must be made." Generally, this is not true. USEPA guidance and practice requires comprehensive site characterization before a remedy decision is made, and thus DNAPL or other complexities are usually well enough defined to predict whether cleanup levels can be achieved or whether a TI waiver makes sense at the onset. In addition, chemical fate and transport science is advanced enough to make reasonable predictions of cleanup times. It is more likely that the problem lies with an unwilling agency or with poor TI waiver requests.

The Department of Defense (DOD) has a major stake in Superfund costs, and it understands the need for TI waivers. DOD has more than 140 NPL (aka Superfund) sites (USEPA 2013b; about 10 % of the total) and most likely a much higher percentage of the total anticipated Superfund costs. DOD's branches offer study reports and guidance on TI waivers (Malcolm Pirnie 2004; NRC 2003; U.S. Air Force 2006).

USEPA's 1995 internal guidance warns that TI reviews will be complex; must be led from the regions; and must consider technical, legal, and policy issues (USEPA 1995). The government TI review team should consist of the regional project manager, a USEPA attorney, groundwater and risk assessment specialists, USEPA Headquarters representatives, and possibly state personnel and a peer reviewer. USEPA's 2011 TI waiver clarification claimed that recent technical improvements make DNAPL cleanup more feasible (i.e., less need for TI waivers), but that TI waivers should still be considered in terms of the agency's original guidance (USEPA 2011). It is unclear if technological improvements have really changed much regarding DNAPL remedy feasibility.

USEPA TI Waiver Guidance

The 1986 Superfund Amendments and Reauthorization Act (SARA) recognized the need for TI waivers (U.S. Congress 1986). USEPA first issued TI waiver guidance in 1993 (USEPA 1993a) and further in 1995 (USEPA 1995), with clarifications in 2011 (USEPA 2011). Although the majority (73 %) of TI waivers have indeed been "front-end" (Malcolm Pirnie 2004), too few TI waivers are granted and the annual rate is declining (USEPA 2012).

The use of TI waivers varies among USEPA Regions, with Regions 1 and 3 (northeast) issuing many more than Region 4 (southeast) or Region 10 (northwest; USEPA 2012). The low number of actual TI waivers represents an apparent skepticism of the approach among USEPA regions, which often appear to have a "wait and see" attitude, meaning that a waiver will only be considered after an initial effort is made first to clean up the contamination. USEPA essentially admits to this in its 1993 TI waiver guidance, which says, "remediation activities can be conducted in phases to achieve interim goals at the outset, while developing a more accurate understanding of the restoration potential..." (USEPA 1993a). The same guidance says, "TI decisions should be made only after interim or full-scale aquifer remediation systems are implemented because often it is difficult to predict..." The "front-end" TI waivers that have been issued may be due to such obvious conditions that the early decisions were inescapable. Translating the USEPA guidance, a TI waiver is likely to be considered only after years of active groundwater remediation and monitoring demonstrate that concentrations have reached an asymptote above drinking water standards.

USEPA's Superfund regulations for hazardous waste site remedy selection, such as Section 300.430 of the NCP, say the agency "expects to return useable groundwaters to their beneficial uses wherever practicable, within a timeframe that is reasonable..." When this is not practicable, USEPA expects to "prevent further migration of the plume, prevent exposure...and evaluate further risk reduction." It then clarifies "impracticable" as "from an engineering perspective" with "cost playing a subordinate role," but the agency puts more emphasis on cost for its own fund-financed actions (USEPA 2008).

USEPA offers the following guidance for evaluating/justifying a TI waiver (USEPA 1993a):

- Identification of the ARAR(s) that need to be relaxed and the geographical area to be covered by the TI waiver
- A conceptual site model with which to consider the TI waiver (essentially, a technical analysis of the contamination)
- An evaluation of restoration potential (i.e., the degree of hope)
- A remedy performance analysis (e.g., design, operation, predicted results, and possible enhancements of alternatives)
- A restoration timeframe analysis ("longer than 100 years" is suggested as "long")
- Cost estimates of existing or proposed remedy options (i.e., to allow consideration of "inordinate costs")

USEPA allows for a TI waiver if the cost of attaining ARARs would be "inordinately high." TI waivers are not a free pass, however. USEPA's guidance makes it clear that in order to obtain a TI waiver, it must be demonstrated that treatment and removal will still occur to the maximum extent possible, remaining sources will be contained, residual groundwater contamination will be managed (e.g., containment of a plume's leading edge), and risks will be minimized (USEPA 1993a).

In 2011, USEPA noted that the mere presence of DNAPL or contamination in fractured bedrock is insufficient justification for a TI waiver (USEPA 2011). This makes sense to a limited degree, because it is sometimes possible to completely remove DNAPL, which would avoid a TI waiver. However, many DNAPL sites, particularly if fractured bedrock is involved, are impossible to clean up, so a good waiver request must prove that. As USEPA (1993b) has recognized, the unique subsurface behavior of DNAPLs makes them "a serious challenge to conventional site investigation and remediation techniques."

More Perspective

Superfund aims to protect people and the environment from hazardous waste contamination. While it is clear that USEPA prefers treatment to management, it is also now obvious that many hazardous waste conditions merit the latter over the former to achieve Superfund's goals. Given what we know technically about the conditions at a site that merit a TI waiver consideration, is it punitive to first require a remedy that will ultimately fail to achieve unrealistic goals? Whether it is taxpayer dollars or industrial profits, there is only so much money to spend on the environment, and we should spend it wisely.

Environmental protection—legacy cleanup or operations compliance—has become a fact of life and a cost of doing business. Most industries accept this relatively new paradigm and try their best to comply. At the same time, most industries do not want to waste money or time. This should translate to Superfund responses that make sense. Would it not be better to achieve Superfund's goals for twice the sites in half the time using realistic goals with appropriate TI waivers? It can be done, but agencies must be willing, and waiver requests must be well justified.

References

Malcolm Pirnie (2004) Technical Impracticability Assessments: Guidelines for Site Applicability and Implementation Phase II Report. Prepared for U.S. Army Environment Center. March

National Research Council (NRC) (2003) Environmental cleanup at navy facilities: adaptive Site Management. National Academies Press, Washington, DC

USEPA (1990) A guide to selecting superfund remedial actions. Directive 9355.0-27FS. April

USEPA (1993a) Guidance for evaluating the technical impracticability of ground-water restoration. Interim Final. Directive 9234. 2–25. September

USEPA (1993b) Evaluation of the likelihood of DNAPL presence at NPL sites. EPA 540R-93-073. September

USEPA (1995) Consistent implementation of the FY 1993 guidance on technical impracticability of ground-water restoration at superfund sites. OSWER Directive 9200. 4–14. January 19

USEPA (1997) Rules of thumb for superfund remedy selection. EPA 540-R-97-013. August

USEPA (1998) Evaluation of subsurface engineered barriers at waste sites. EPA 542-R-98-005. August

USEPA (1999) Use of monitored natural attenuation at superfund, RCRA corrective action, and underground storage tank sites. Directive 9200. 4–17 pp April 21

USEPA (2003) The DNAPL remediation challenge: is there a case for source depletion? (by Expert
    Panel on DNAPL Remediation) December
USEPA (2008) 40 CFR Part 300 – National oil and hazardous substances contingency plan (at 40
    CFR Section 300.430)
USEPA (2011) Memorandum re: clarification of OSWER's 1995 technical impracticability waiver
    policy. OSWER Directive 9355. 5–32. September
USEPA (2012) Summary of technical impracticability waivers at national priorities list sites. OS-
    WER Directive 9230. 2–24. August
USEPA (2013a) National priorities list website. May 24. http://www.epa.gov/superfund/sites/npl/.
    Accessed 1 Aug 2013
USEPA (2013b) National priorities list (NPL) sites by agency website. May 31. http://www.epa.
    gov/fedfac/ff/nplagency2.htm. Accessed 1 Aug 2013
U.S. Air Force (2006) Incorporating technical impracticability into the air force cleanup program.
    March 22
U.S. Congress (1986) Superfund amendments and reauthorization Act of 1986. Public Law 99–
    499. October 17

Chapter 13
Superduperfund

Abstract Many Superfund site responses excavate wastes and ship them off-site for landfilling. In such cases, Resource Conservation and Recovery Act (RCRA) rules apply, and the wastes must be rendered nonhazardous prior to landfilling. As long as RCRA rules apply for those off-site landfills, and as long as the nonhazardous rendering remains effective, there will be no problem with Superfund's off-site shipments. Otherwise, we will face Superduperfund sites in the future.

Keywords RCRA · Land ban · LDR · Rendering · Disposal · Institutional controls · NPL · Subtitle C · Subtitle D

Introduction

Congress passed Superfund (Comprehensive Environmental Response, Compensation, and Liability Act, CERCLA) in 1980. Since then, this federal program has addressed (cleaned up or underway) about 1,700 hazardous waste sites. With state and federal programs combined, there are about 300,000 hazardous waste sites in the USA. Treatment, removal, and containment are the common cleanup approaches. For example, contaminated soils might be removed, or they might be contained with a subsurface slurry wall around the sides and a cap over the top.

Removal versus containment often depends on cost and anticipated future property use. For example, if residential redevelopment is anticipated, removal is often preferred to ensure safe installation of basements. When contaminated soils are removed, they are often disposed elsewhere (sometimes they are treated and redeposited on site). When disposed elsewhere, Resource Conservation and Recovery Act (RCRA) rules apply, meaning soils will be manifested, transported, treated, and landfilled in a licensed facility.

CERCLA deals with legacy hazardous waste issues, while RCRA deals with hazardous waste management from active facilities, such as from factories and in landfills. Together, they address nearly all hazardous waste issues, although other programs, such as the Clean Water Act (CWA), have waste treatment requirements that might also be considered hazardous waste management.

N. Shifrin, *Environmental Perspectives,* SpringerBriefs in Environmental Science, 85
DOI 10.1007/978-3-319-06278-5_13, © The Author(s) 2014

The neighbors around a Superfund site breathe a sigh of relief when wastes are carted away. But will the ultimate disposal sites eventually become Superduperfund sites? The RCRA sites receiving Superfund wastes are designed to safely contain hazardous wastes and are currently monitored to confirm that. In addition, licensed RCRA disposal facilities have rigid closure and post-closure care requirements. But will their owners remain viable for eternity, and will the control systems remain functional beyond the 100 years of economic present-value infinity? Of course not. To determine if we are simply kicking the can down the road, it is useful to consider the nature of present Superfund remedies and their RCRA elements.

Superfund Remedy Statistics

About 40 % of Superfund sites use off-site disposal as a part of the overall remedy (USEPA 2010).[1] This statistic is complex, however, because it reflects only off-site disposal use, not the amount, which available information defies defining. About 57 % of Superfund sites have remedies involving no treatment (e.g., institutional controls, caps, walls, and off-site disposal); thus, 43 % include treatment, albeit sometimes partial. About one third of the "no treatment" remedies and half of the "treatment" remedies use off-site disposal. This implies that Superfund wastes sent for off-site disposal are not treated by Superfund about half the time. A common pretreatment for off-site disposal is solidification/stabilization.

Superfund source control remedies are predominated by solidification/stabilization (21 %, on-site and off-site), followed by in situ soil vapor extraction (14 %), thermal desorption and incineration (8 %), chemical treatment (5 %), and a number of other approaches such as bioremediation, soil flushing, and phytoremediation.

Groundwater remedies at Superfund sites almost always rely on institutional controls (94 % in 2008, up from 15 % in 1995), with the formerly popular pump-and-treat remedies (90 % in the 1980s) down to 25 % in the 2000s. In situ treatment (e.g., oxidant injection, air sparging, or permeable reactive barriers) and monitored natural attenuation have each grown to about 30 % in the 2000s. Only 8 % of the Superfund groundwater remedies in 2008 relied on treating groundwater at the supply end.

As of 2013, the NPL contained 1,320 Superfund sites, and 365 additional sites had been delisted/completed. One estimate is that about US$ 200 billion will be spent to clean up 294,000 hazardous waste sites in the USA (USEPA 2004). USEPA clearly favors treatment over removal/disposal, but the latter still occurs at a significant number of Superfund sites.

---

[1] These statistics are through 2008.

RCRA Requirements

RCRA was passed in 1976 to manage wastes from operating facilities and manage ongoing disposal sites, which were categorized by their description in the law as Subtitle C (hazardous) or Subtitle D (nonhazardous). RCRA wastes are either listed (currently more than 500 are on the list) or "characteristically hazardous," as determined by testing. Hazardous characteristics are ignitibility, corrosivity, reactivity, or toxicity. The latter is tested for 40 compounds, with defined leachate thresholds by the Toxicity Characteristic Leaching Test (TCLP). If a waste is either listed or characteristically hazardous, it must be rendered nonhazardous under RCRA's Land Disposal Restriction (LDR) rules before placement into a Subtitle C landfill. For example, soils failing the LDR for benzene might be rendered nonhazardous by air sparging to reduce benzene concentrations. Otherwise, it is a "solid waste" and can be placed in a Subtitle D landfill. Operating landfills are licensed under RCRA (or state equivalents) and are required to be monitored for leakage, with careful bookkeeping about what and how much is landfilled.

Typical Subtitle C landfills have double liners at their bottom and sides with leachate detection and collection systems between these liners. When a Subtitle C landfill is full, it must be closed with a cap and monitored for the foreseeable future, according to closure and post-closure care plans submitted by the owner and approved by the agency.

RCRA's LDR program prohibits the land disposal of untreated hazardous wastes, and dilution is not considered treatment (USEPA 2001). Depending on the waste/chemicals, LDRs can be either concentration limits or treatment technology specifications. Although there are exceptions (hazardous fluids disposed under the CWA, for example), once treatment is required for a particular hazardous constituent, "universal treatment standards" apply, which means that all other hazardous constituents in the waste must also meet these regulatory thresholds.

In theory, LDRs provide a safety net for Subtitle C landfills when, at some point in the future, their physical controls deteriorate or the owners disappear and post-closure care is in doubt. However, there are four chinks in the armor: (1) the LDR program was phased in starting circa 1984, so earlier wastes were untreated; (2) newly restricted wastes are assigned effective dates, and they could be landfilled untreated until those dates; (3) LDRs do not apply to on-site RCRA solutions (e.g., AOCs and CAMUs); and (4) the long-term effectiveness of LDR treatment requirements has not been tested.

At the present time, there are 21 licensed Subtitle C landfills in the USA (Environment Health and Safety Online 2013) to receive the 2 million tons per year of RCRA hazardous waste being landfilled (USEPA 2008, 2011). Some of this hazardous waste is from Superfund sites. Although Superfund reports some of that waste as untreated (USEPA 2010), it most likely is treated later under the LDR program as it gets deposited in a RCRA landfill.

When RCRA Meets CERCLA

As noted above, many Superfund sites dispose of wastes/soils in RCRA landfills. Although on their surface RCRA regulations offer tight controls on landfill safety for hazardous wastes—pretreatment, double liners, caps, monitoring, and long-term maintenance—it is still simply a matter of time for some or all those precautions to fail. What will happen when they do? This is not just a Superfund issue, but Superfund wastes will play a role.

The RCRA program, including LDRs, is sensible, given that our production of hazardous wastes is unavoidable today. But today's hazardous landfills are sizeable—1 million tons each at the current rate if each landfill operates for 10 years. Eventually the liners and cap will degrade, and the long-term care party will disappear, resulting in a new order of hazardous waste problems—unless the LDR does its job. In many cases, the LDRs do not destroy hazardous chemicals; they simply render the waste to either reduce its mobility or contaminant concentrations. The unanswered question therefore is whether the rendered hazardous waste in those large, concentrated zones will eventually pose new, larger subsurface contamination issues. If not, LDRs are the permanent solution; if so, we will be faced with Superduperfund. In addition, some Subtitle D landfills may offer future hazardous chemical releases.

One way Superfund might minimize its role in this future is to require that any waste/soil shipped for off-site disposal be treated beyond LDR requirements, if possible. Regardless, we now have more than 30 years of experience cleaning up and managing hazardous wastes, with improving technology and environmental science to some degree, so perhaps it is time to take stock. We were caught flat-footed in the first wave, which resulted in Superfund; we now should ensure our heirs will not face Superduperfund.

References

Environment Health and Safety Online (2013) Commercial hazardous waste landfills. http://www.ehso.com/cssepa/tsdflandfills.php. Accessed 7 Aug 2013

USEPA (2001) Land disposal restrictions: summary of requirements. EPA 530-R-01–007

USEPA (2004) Cleaning up the nation's waste sites: markets and technology trends. EPA 542-R-04–015

USEPA (2008) EPA's report on the environment. EPA/600/R-07/045F. http://cfpub.epa.gov/eroe/index.cfm?fuseaction=detail.viewPDF&ch=48&1ShowInd=0&subtop=228&lv=list.listByChapter&r=239786. Accessed 7 Aug 2013

USEPA (2010) Superfund Remedy Report, thirteenth edn. EPA-542-R-10-004. Office of Solid Waste and Emergency Response (OSWER)

USEPA (2011) Quantity of RCRA hazardous waste generated and managed. http://cfpub.epa.gov/eroe/index.cfm?fuseaction=detail.viewPDF&ch=48&1ShowInd=0&subtop=228&lv=list.listByChapter&r=239786. Accessed 7 Aug 2013

Chapter 14
Heads in the Sand

Abstract The "not in my backyard" (NIMBY) syndrome prevails for many proposed projects, such as new landfills or new energy projects. Unfortunately, we need these new projects more than ever. Assuming that a proposed project has been designed to minimize actual risks, only perceived risks remain. But those perceived risks can still stymie a proposed project. A more rational approach to spending years fighting the maze of obstacles thrown in front of such projects would be to offer clear benefits to offset the perceived risks, right from the beginning, as part of the project design.

Keywords NIMBY · Siting · Risk

Introduction

Almost no one wants to be near waste or energy projects. Every now and then, an accident occurs and our fears are fueled. Green efforts and government regulations can reduce these fears, but not entirely. For example, our 104 nuclear reactors (USEIA 2012b) continue to generate waste without a viable long-term plan because we cannot agree on where to send the waste. No amount of denial will reduce our hazardous waste management issues. The solution lies in dispelling the myths about the hazards, developing appropriate safeguards, and offering appropriate compensating benefits for allowing these "hot potatoes" in our backyards.

The Brutal Facts

By the end of the twentieth century, the USA had essentially abandoned major hazardous waste destruction by incineration and alternative energy generation by burning wastes because of siting hurdles. The term "NIMBY" (not in my back yard) entered our lexicon as a reverberating echo from Love Canal. Chemical hazardous waste today is managed under the 1976 Resource Conservation and Recovery Act (RCRA) and the 1980 Comprehensive Environmental Response Compensation and

N. Shifrin, *Environmental Perspectives,* SpringerBriefs in Environmental Science,   89
DOI 10.1007/978-3-319-06278-5_14, © The Author(s) 2014

Liability Act (CERCLA) for active and abandoned chemical sites, respectively. RCRA regulated about 25 million t of such hazardous waste in 2009 from 175,000 generators (USEPA 2011a). RCRA also aims to clean up legacy hazardous waste problems at factories ("corrective actions") and oversees 4,000 such actions (USE-PA 2013a). CERCLA addresses more than 1,300 hazardous waste sites (USEPA 2013b) ranging in complexity and size from old landfills to the Hudson River.

Radioactive waste also surrounds us. The proposed Yucca Mountain nuclear waste repository in Nevada is perhaps the best example of NIMBY's unfortunate outcomes. Yucca Mountain was intended to receive ultimately 70,000 t of nuclear waste (USDOE 2002), 90 % from power plants and 10 % other high-level waste (USEPA 2001), at a rate of 3,000 t per year for 24 years (Eureka County, Nevada, Nuclear Waste Office 2013) at a cost of almost US$ 60 billion (USDOE 2002). The US government spent more than US$ 14 billion for research and preliminary site preparations (GAO 2009). The government used a tunnel machine to create a 5-mile tunnel 1,000 ft below the ground and 1,000 ft above the water table so it could study the proposed design (USEPA 2001). Geologists mapped essentially every crack in the bedrock surface of that tunnel. They took seismic and many other measurements. They designed secure storage canisters for the waste and storage compartments for the canisters that ultimately would occupy 100 miles (USEPA 2001) in a dry, stable subsurface that would protect the waste from release and human exposure for thousands of years. They designed a transport system to move nuclear waste to the storage facility that could withstand train collisions without releasing waste.

In the end, however, the project was abandoned, and one reason given was the site's potential instability—Yucca Mountain was located near a 10,000-year-old volcano site. Between the lines, there was political pressure to abandon the project, including Nevada's opposition. So the solution remains today as no solution—we continue to be surrounded by less than optimal radioactive waste storage.

At the opposite end of the spectrum is the Hudson River polychlorinated biphenyl (PCB) sediment dredging example. An estimated 2.65 million cubic yards of PCB-laden sediments are being dredged from the upper Hudson River, with the dewatered sediments being shipped to Andrews, TX, for deposit in a landfill constructed there for that purpose (USEPA 2002).

We have many reasons for implementing potentially controversial projects. The USA had 6,600 power plants in 2011, 66 of them nuclear (USEIA 2013). Our energy demand has risen to 97 quads (quadrillion BTUs) from about 78 quads in 1980 (USEIA 2012a), so we need even more power plants. Renewable energy supplies only 9 % of our demand (USEIA 2012a), but even renewable energy projects are locally resisted, such as the windmill project in Nantucket Sound. There are 1,900 municipal (USEPA 2011b) and 21 hazardous waste (USACOE 2006) landfills in the USA. We generate 250 million t per year of municipal solid waste (USEPA 2011b) and 22 million t of hazardous waste (USEPA 2011a) per year (although about 90 % of that is injected into deep wells). The number of landfills is decreasing, the remaining ones are getting larger (up to 9,200 t per day of waste; Parten 2010), and new ones will be needed. How easy will it be to site the next mega-landfill when the current ones are full and our 250 million t per year of municipal waste starts piling up?

We treat 32 billion gallons of wastewater every year in POTWs (Ellis 2004). Luckily, we have learned to make wastewater treatment plants aesthetically pleasing to the neighbors, at least from beyond the fence, and we know how to control the smell. Implementing wastewater treatment projects seems to be the simplest of today's siting issues.

The brutal fact is that we generate astounding quantities of waste and have massive energy and natural resource demands. In general, these quantities are growing along with our population. The latter fact confounds the former—with less "invisible space," our siting problems have become more difficult.

Dispelling the Myth

To some degree, the risks of our waste and energy projects are a myth (i.e., they are *perceived* as a risk but pose very little *actual* risk). In fact, it is likely that our home cleaning fluids and personal care products pose more actual health risk than a landfill next door. Property value diminution is an actual risk but it is usually based on mere perceived health risks. It would appear that clarifying the actual health risks of controversial projects might help with project siting.

The problem with clarifying actual health risks is that people do not believe the analyses, because they have little experience quantifying risk. Although people weigh risks against benefits every day (flying is worth the risk of crashing, for example), they are not used to seeing a risk "number" (i.e., a quantification of risk). For controversial projects, the risk quantification, along with its underlying assumptions, is presented with a conclusion like, "See, it's safe." First, people do not know how to benchmark the "number." Second, people will challenge the assumptions because their experiences are different from them. Finally, the mere quantitation of risk is alien to most laypeople.

An example of the alien nature of the analysis would be an air model for a proposed power plant. The model may be accurate, having been well designed, calibrated, and verified, but most people do not know what an air model is, so why would they believe its predictions?

The possibility that controversial projects are unsafe is most often a myth, but the analyses themselves create fear, and fear creates property diminution. In addition, people weigh risks against benefits, and the benefits presented for controversial projects are often disconnected from the risk takers. Finally, people fear an unanticipated accident, which most often is not part of the risk analysis.

Rational Policy

Both sides of the NIMBY argument, as well as the public in general, could benefit from a more rational siting policy. NIMBY attitudes stem from concern about risks both real and perceived. Project promoters must offer the local risk takers appropriate benefits that outweigh the risks. The cost of controversial projects should include provision of these direct compensatory benefits. Such benefits are not bribes; they are simply a necessary part of the project from the risk taker's view.

An example of an appropriate compensatory benefit might be a neighborhood improvement that offsets potential property diminution. Another might be a health center with free blood monitoring to residents concerned about contamination. Both examples assume the project is indeed safe and that a reasonable effort has been made to prove its safety.

In the Yucca Mountain example, the US public's need for a nuclear repository was not an appropriate benefit for Nevada. Why should they "bear the brunt"? "What's in it for them?" The project promoter must put something in the project that makes local sense in order for locals to agree to it.

Project promoters and the government need to offer a more personal risk–benefit proposition. Currently, the government views a siting proposal in two ways: meeting regulations and the public good. But the locals view it in terms of what is good for *them*. Offering a more general public benefit can keep the government from killing a project, but local opposition can still make a project die of old age. Avoiding failure is not the same as success, and success will come only when locals support a project because they see benefit in it for them.

References

Ellis T (2004) Chemistry of Wastewater, Wastewater Quantities. http://www.eolss.net/EolssSampleChapters/C06/E6-13-04-05/E6-13-04-05-TXT-05.aspx. Accessed: 31 July 2013

Eureka County, Nevada, Nuclear Waste Office (2013) Yucca Mountain.org FAQs. http://www.yuccamountain.org/faq.htm. Accessed: 11 July 2013

Parten C (2010) America's Largest Landfills. CNBC.com. http://www.cnbc.com/id/39382002/page/1. Accessed: 31 July 2013

USACOE (2006) Commercial Hazardous Waste Landfills. http://www.environmental.usace.army.mil/library/pubs/tsdf/sec2-4/sec2-4.html. Accessed: 5 Aug 2013

U.S. Department of Energy (USDOE) (2002) Final environmental impact statement for a geologic repository for the disposal of spent nuclear fuel and high-level radioactive waste at Yucca Mountain, Nye County, Nevada. DOE/EIS-0250. February

U.S. Energy Information Administration (USEIA) (2012a) Annual Energy Review 2011. DOE/EIA-0384(2011). September. Figure 1.1

USEIA (2012b) Energy in brief: what is the status of the U.S. nuclear industry? http://www.eia.gov/energy_in_brief/article/nuclear_industry.cfm. Accessed: 16 July 2013

USEIA (2013) Electric Power Annual 2011. January. Table 4.1

USEPA (2001) A reporter's guide to Yucca Mountain

USEPA (2002) Hudson River PCBs site, New York, Record of Decision

USEPA (2011a) Quantity of RCRA hazardous waste generated and managed

USEPA (2011b) Municipal solid waste in the United States. 2011 facts and figures. EPA530-R-13-001. May

USEPA (2012) Summary of technical impracticability waivers at national priorities list sites. OSWER Directive 9230. 2–24. August

USEPA (2013a) Learn about RCRA corrective action. http://epa.gov/epawaste/hazard/corrective-action/learn.htm. Accessed: 23 July 2013

USEPA (2013b) National Priorities List (NPL). http://www.epa.gov/superfund/sites/npl/. Accessed: 7 Aug 2013

U.S. Government Accountability Office (2009) Report to congressional requesters. Nuclear waste management. Key attributes, challenges, and costs for the Yucca Mountain repository and two potential alternatives. GAO-10-48. November

Chapter 15
The Safety of Chemical Products

Abstract Exposures to everyday products such as household cleaners and personal care products potentially pose more actual exposure to chemicals than do Superfund sites. Product liability suits instill incentives for manufacturers to create safer products, but for those specific cases, the damage has already been done. A much-needed supplement to such indirect incentives is a product safety rating system. The system would need to be simple and clear. It would provide consumers with valuable information beyond today's typical price–performance paradigm, and it would provide manufacturers a marketing tool with the creation of safer products as a benefit to all.

Keywords Product safety · Product liability · Precautionary principle · Enterprise risk management · MSDS · Toxicity profile · Rating

Introduction

Many consumer products contain chemicals that might offer unhealthy exposures. The health risk of such products depends on the actual exposure, which involves how a product is used or misused, in addition to the product's potential for chemical emission. Household cleaners, plastics, flame retardants, pesticides, and personal care products are a few items that have the potential for health risks.

There are at least three ways to address potential health risks from products: (1) remove their toxic compounds; (2) provide information to allow informed buying and use decisions; and (3) create negative incentives, such as by lawsuits after harm is caused. McDonough and Braungart (2002) made a case for removing toxic chemicals from products. Post-sale product liability and defense considerations are discussed below. In addition, a rating system for buying/use decisions is discussed below.

It might be argued that the "precautionary principle" should apply to the sale of chemical products. The principle argues for prudence in the face of potential risk when consensus has not yet been reached on the actuality of that risk. Because chemical products might offer such a wide range of risk, including no risk in some

Robin Cantor, Ph.D., of Berkeley Research Group's Washington, DC, office coauthored this chapter.

cases, a product safety rating system is suggested to represent such prudence in conformance with the precautionary principle.

Product Liability

In recent years, product liability and product safety have become high-priority issues for policy makers and business decisions (Cantor 2012). The application of damages from product liability can be viewed as either punishment or the provision of incentives for safer products. The tension in business communities between demands for increased product safety and decreased liability serve to highlight the complexity of the issues that corporate risk managers, policy makers, business advisors, and consumers wrestle with on a daily basis to manage product risks.

Product liability is an important area of the US legal and regulatory systems and is important in the context of enterprise risk management (ERM). Modern ERM emphasizes a proactive product liability focus in a world with many different types of risks. For that reason, ERM has inexorable links with the incentives and penalties inherent in the legal and regulatory structures under which enterprises operate. Product liability theories and standards have evolved from a traditional foundation in specific and demonstrated manufacturer causation to more complex theories of market, successor, and other "controller" liabilities. In turn, this has expanded the set of issues surrounding questions such as *Who absorbs the cost of damages, and over what time period?* In this section, the expanding scope of product liability issues in the context of ERM and post-sale liability are addressed.

In the USA, incentives to limit or manage product liability are most typically associated with tort litigation. *Product liability* is a general term that applies to several possible causes of a compensable injury, traditionally including negligence, breach of warranty, and fraud. While these causes continue today, contemporary product liability litigation often focuses on causes based on defect considerations and claims about manufacturer conduct. Importantly, strict liability for known or reasonably known product defects can attach to the manufacturer even if there is no aspect of negligence or improper conduct. Not surprisingly, this legal view is controversial from a law and economics perspective that emphasizes efficiency and maximizing innovation and product development (Polinsky et al. 2010).

Modern product liability legal standards distinguish three categories of product defects:

1. Manufacturing defects: When the product departs from its intended design, regardless of the level of care exercised by the manufacturer.
1. Design defects: When the reasonably known risks of harm posed by the product could have been reduced or avoided by the adoption of safer commercial technologies or product alternatives.
2. Inadequate instructions or warnings defects: When the reasonably known risks of product-related injuries could have been reduced or avoided by reasonable instructions, labels, or warnings (ALI 1998).

Damages paid to injured parties are essentially the ex post imposition of an increase on the cost of production. In the post-sale world of products, however, production might have occurred long ago and possibly involved parties that no longer exist. Even when production is not in the distant past, the (explicit or implicit) economic agreements in the supply chain often do not specify how unanticipated production costs should be allocated among the parties. This is not surprising, given that the mitigation costs of the risks may have been unanticipated at the time these contracts and economic relationships were active. Nonetheless, science and technology have facilitated our capabilities to identify and measure exposures to potentially harmful substances in products that enter the stream of commerce. Sometimes these exposures are associated with potential injuries for which it can be difficult or impossible to identify the parties responsible for the harmful products. Examples of these types of exposures occur with pharmaceutical products, water contamination, air pollution, and product additives widely used in many applications.

Under traditional liability theories, the determination of many aspects of causation is complicated by requirements for evidence, assignment of responsible actions by the various parties, the limits of scientific knowledge, and the manifestation of injuries. The legal system sometimes provides an alternative mechanism for compensation for exposure to these risks. Courts have considered alternative, although controversial, theories of product liability that have been and are based on industry data and market participation of the suppliers, rather than on specific proof about the individual conduct of a particular manufacturer.

In addition, product liability is not based solely on defects known at the time of sale, but can also attach to any part in the supply and distribution chain if the product defect reasonably should have been known to the controlling party. Other considerations removed from the specific time of sale might include liability for post-sale failure to warn and successor liability.

Whether the threat of tort actions provides sufficient private incentive for manufacturers to have effective monitoring, product recall, and risk management strategies in place are an important social issue. For example, rulings in a number of prominent lawsuits relating to off-label usage and promotion of drugs indicate that the onus rests with the manufacturer for proper labeling and disseminating information relating to drug usage and potential harms (Welt and Anderson 2009). In this respect, measures such as supplier verification, approval programs, and informative labeling may facilitate effective reductions in product liability risk faced by manufacturers.

Going forward, science will continue to change the need for ERM to monitor diligently for future consequences and sources of product liability. Genotoxicity is an example of an emerging area in risk monitoring. This area builds on research that suggests that exposure to chemicals during embryonic development can result in DNA changes that might lead to toxic tort allegations. From a manufacturer's perspective, this introduces a new outlook on legacy liabilities that may persist after the product has been discontinued for generations.

Given the substantial costs imposed by tort proceedings, parties in market transactions often attempt to limit their liability for injuries by adding either warning labels to products or liability-release provisions to sale, rental, or licensing contracts. Some commentators have noted that when a pre-contractual relationship exists between sellers and consumers, tort damages are a socially inefficient means of internalizing the costs of accidents or injuries (Rubin 1993). Ultimately, product prices reflect tort damages. According to this view, consumers and society would be better off if buyers and sellers negotiated directly to determine the sharing of risk for defective products based on which party can most effectively bear the risk. This view supports the use of limited product warranties, product labeling, and liability release provisions in contracts or at the point of sale, to lower product costs and encourage the use of risky but beneficial products.

Product Safety Rating

A potential enhancement for ERM would be a chemical product rating system. Such a system would provide manufacturers with a rational way to consider their products, while providing consumers with pertinent information for buying and use decisions. An effective system would not necessarily eliminate products with lower ratings, because those products might still have large enough benefits to remain viable. Both risks and benefits are important.

A useful rating system would consolidate elements scattered through various government efforts in existence today, but should be a voluntary system that is simple to understand. The issues involved with product safety can be incredibly complex, but only a simple rating system will be successful. Effectiveness depends on many factors, but two conditions worth considering are ease of use and the ability to complete the user's understanding of risks. Perhaps it is best to split the rating system into two parts: a summary rating and easy access to supplemental information. The challenge for development of a rating system will be to capture the important issues while packaging it in an understandable and meaningful way—simplistic elegance, not simplistic ignorance.

Easy access to supplemental information may allow for a simpler summary rating system if it allows users to expand their understanding of the ratings, compare their exposures to that assumed by the rating, and compare across products for purchase and use decisions. Such easy access in today's Internet and smartphone world is practical and provides an interesting Silicon Valley project. But delivery is only half the challenge; the key to product rating success will be to make the information meaningful. For example, much pertinent information exists—such as in US Environmental Protection Agency (USEPA) pesticide registration material, materials safety data sheets (MSDS), USEPA's toxicity database (IRIS), or the US Centers for Disease Control's toxicity profiles—but this information is not meaningful to consumers. It is not reasonable to expect consumers to sift through LD50s (lethal dose 50% of the time) in toxicity profiles and create their own method of comparison. Meaningful simplification will be challenging because it will involve notions of toxicity, exposure, and risk communication—but it is necessary.

The mere presence of a toxic compound in a product is only part of the story. Consumers must also understand if their use of a product offers higher exposures than what is assumed by the rating system, in case they want to seek a safer-rated product for a higher exposure pattern. Thus, easy access to the assumed exposure scenario for each product rating would be useful (e.g., shampoo used once a day in a 5-minute shower). It may be that consumers wash their hair twice a day, but a safer-rated product does not wash as well so the benefits of the less safe product outweigh the risks. In a market economy, consumers should have the right information to make that decision. Importantly, empowering consumers with accessible information has effectively and substantially changed their relationship in other markets, such as health care and energy, and it could do the same for consumer products.

It will not be easy to create an effective chemical product rating system and meaningful corresponding supplemental information. However, such a system could satisfy the consumer on many levels and could be used by manufacturers to develop safer products with marketing advantages.

Product Gestalt

While reactive product liability responses have a place after damage has been done, it is also possible to do a better job preventing harm in the first place. Such a proactive stance involves better product design and a simple rating system with straightforward access to additional and pertinent information. Manufacturers certainly want to make products safer, and much progress is being made. Clearly, the products we use have benefits, but consumers need a new lexicon to weigh those benefits against potential risks.

References

American Law Institute (ALI) (1998) Restatement of the Law Third, Torts: Product Liability, 382 pp
Cantor R (2012) Product liability. In: Bainbridge WS (ed) Leadership in science and technology: a reference handbook. Sage Publications Inc., Thousand Oaks, pp 281–288
McDonough W, Braungart M (2002) Cradle to cradle, remaking the way we make things. North Point Press, New York, 193 pp
Polinsky A, Shavell M, Shavell S (2010) The uneasy case for product liability. Harv Law Rev 123(6):1437–1492
Rubin PH (1993) Tort reform by contract. AEI Press, Washington, DC, 91 pp
Welt MM, Anderson EL (2009) Changing perspectives on chemical product risks. John Liner Rev 23(3):58–72

Index

95% upper confidence limit (95UCL), 27, 51

A
Accuracy, 19, 20, 29, 36, 66
Advection, 44
Air models (AEROMOD), 46
Air quality, 25, 63, 64
    regulation, 67–69
    sources, 65
Analytical methods, 20, 21, 28, 32, 34

B
Baghouse, 67
Bar chart, 37
Biochemical oxygen demand (BOD), 45, 56, 58, 59

C
Calibration, 22, 44, 47
Cancer slope factor (CSF), 50, 52
Carrying capacity, 1
Cause and effect, 9, 34, 43, 45
Chemical species, 52
Classification, 55, 56, 61
    stream, 11, 12
Composite samples, 26
Containment, 80, 85

D
Data
    posting, 37
    quality objectives, 26
Database, 35, 36
Dense non-aqueous phase liquids (DNAPL), 79–82
Detection limit, 19–21, 23, 36

Differentiation, 31
Dispersion
    coefficients, 44
Disposal, 8, 9, 56, 86
    of wastes, 7–9, 12

E
Electrostatic precipitator (ESPs), 67
Emissions, 68
    carbon compound, 64
Enterprise risk management (ERM), 96–98
Environmental
    forensics, 31, 34
    regulation, 1, 11
Exposure point concentration (EPCs), 51

F
Fugitive emissions, 44, 46, 65, 66

G
Gas chromatography (GC), 22
Gaussian plume, 44
Geographic Information Systems (GIS), 36

H
Hazard index (HI), 50, 53
Hazardous air pollutants (HAPs), 64, 65
High-volume (HiVoL), 64
Hydrocarbon fingerprinting, 32

I
Institutional controls, 86
Integrated Risk Information System (IRIS), 52, 98
Introduction, 2, 14, 53
Isoconcentration contours, 36
Isotopes, 32